Otto Stoll, Frederick Du Cane Godman, Osbert Salvin

Biologia Centrali-Americana

Otto Stoll, Frederick Du Cane Godman, Osbert Salvin

Biologia Centrali-Americana

ISBN/EAN: 9783741132650

Manufactured in Europe, USA, Canada, Australia, Japa

Cover: Foto ©Klaus-Uwe Gerhardt /pixelio.de

Manufactured and distributed by brebook publishing software
(www.brebook.com)

Otto Stoll, Frederick Du Cane Godman, Osbert Salvin

Biologia Centrali-Americana

BIOLOGIA
CENTRALI-AMERICANA.

ARACHNIDA ACARIDEA.

BY

Prof. OTTO STOLL, M.D.

1886-1893.

CONTENTS.

INTRODUCTION.

Compared with the work that has been done in Europe upon the Arachnida Acaridea from the time of the classical writers, Linnæus, De Geer, Hermann, and Latreille, up to that of Dugès, Koch, Nicolet, P. Kramer, Mégnin, Michael, Neuman, Berlese, Canestrini, Nalepa, and many others, the American literature of this group of animals is exceedingly scanty. Thomas Say (1821), one of the first entomologists in the United States, paid some attention to this neglected group. Later on (1836), Dana and Whelpley, as well as Haldeman (1842), described and figured some North-American species of Hydrachnidæ. In our times we meet with the well-known names of Riley and Packard in connection with North-American Acarids of various groups, and Mr. Harry Garman has published a paper on the Phytoptidæ. In 1866, Messrs. Herbert Osborn and Lucien M. Underwood gave a " Preliminary List of the Species of Acarina of North America " in the 'Canadian Entomologist.' In Mexico, M. Alfred Dugès has published several valuable articles on various species of Acarids inhabiting that country, and N. Conil in Buenos Ayres has done the same for some species of the Argentine Republic.

A certain number of American Acaridea have been described by European authors, and some of the larger and more conspicuous species were already known to the founders of Entomology, to Linnæus, De Geer, and Fabricius. In modern times C. L. Koch has described several Ixodidæ from various parts of America, and Trouessart and Mégnin have studied some forms of Dermaleichidæ which live on American birds. To Antonio Berlese and R. Canestrini we are indebted for some most valuable papers on Acaridea from the La Plata region and from Brazil.

But nevertheless we are far from possessing a knowledge of the American Acarid fauna comparable to that which we have long had of the European forms.

In the following pages an attempt has been made to fill, to some extent, the gap which at present separates the forms known from North America and those described by

the above-quoted Italian authors from various parts of South America, and to charac-
terise some new types from Mexico and Central America. The materials which I had
at my disposal were derived from various sources. During a stay of nearly five years
in various parts of Guatemala, I tried to make myself acquainted with the Acarid-
fauna of that country, by studying, as far as the unsettled life of a medical practitioner
would allow, the living forms. I had then with me a Hartnack microscope of but
moderate powers (objectives 4 and 7), the lenses of which became damaged by the long
influence of the excessive humidity of the climate of the Costa Grande. Moreover, I
laboured under an almost total want of modern literature on Acarids, having only some
of the works of the older writers with me. These unfavourable circumstances will
account to some extent for the differences the reader may find in the execution of the
drawings and the descriptions of some species, several of which are too delicate for
preservation in alcohol, the only method then within my reach. Many of my drawings,
especially of Gamasidæ, have been entirely omitted from the present memoir, as being
too incomplete to allow a comparison with the European forms; and even amongst
those which I have admitted there are some which I should have liked to revise again
from the specimens themselves.

Fortunately, this lack of preserved materials has in numerous cases been made up by
dried specimens obtained by other naturalists who have travelled in Central America.

The fact that other collectors have frequently fallen in with the same species as myself
shows that my researches, though far from complete, were sufficient at least to give an
idea of the composition of the Acarid fauna of a tropical country, and to enable me to
come to some general conclusions as to the geographical distribution of the various
groups.

The most striking fact elicited by the study of the Acarid fauna of Central America
is undoubtedly the great similarity between the types with which we are familiar in
the temperate regions north of the European Alps and those occurring in the gloomy
shadow of the tropical forest. Under the bark and in the fissures of putrefying tree-
trunks in the tropics we meet with some minute Gamasidæ which only by a close
microscopical examination can be distinguished from European species living under
similar conditions. Upon various beetles of the families Copridæ and Passalidæ are
found, with others, such well-known forms as *Gamasus* (*Holostaspis*) *marginatus* and
G. coleoptratorum, species already described by authors of the past century. Among
the fallen leaves and in the decaying fruits of *Theobroma*, *Lucuma*, and other tropical

trees, which lie scattered over the damp ground of the forest, we frequently observe the slow-moving Oribatidæ, or the soft velvety Trombidiidæ, the larvæ of which may be found attached to the wings of dragon-flies, grasshoppers, and other insects. The larger species of Muscidæ carry on their thorax the hypopial forms of Tyroglyphidæ, just as they do in Europe; and the Belostomidæ, which we find in the ponds of the high valleys and barrancas of Guatemala, have their legs infested by the larvæ of Hydrachnidæ, in the same way as are those of *Ranatra* and *Nepa* in Europe. It is a remarkable circumstance that not one of the species described in the following memoir represents a generic type entirely new or peculiar to Central America. With the exception of the holotropical genus *Megisthanus* (fam. Gamasidæ), and the chiefly American genus *Amblyomma* (fam. Ixodidæ), all the genera enumerated by me likewise occur in the Palæarctic Region.

Geographical Distribution of the Genera of Acarida hitherto found in Central America.

Fam. TROMBIDIDÆ.

Gen. TROMBIDIUM.
    Palæarctic Region (Europe, N.E. Siberia).
    Nearctic R. (U.S.A.).
    Neotropical R. (Central and South America).
    Æthiopian R. (Senegambia, Gold Coast, S.W. Africa, Cape of Good Hope).

Gen. RHYNCHOLOPHUS.
    Palæarctic Region (Europe).
    Nearctic R. (U.S.A.).
    Neotropical R. (Central and South America).

Gen. LIMNOCHARES.
    Palæarctic Region (Europe).
    Neotropical R. (Guatemala, Buenos Ayres).

Fam. ACTINEDIDÆ.

Gen. ACTINEDA.
    Palæarctic Region (Europe, Siberia).
    Neotropical R. (Central and South America).

Fam. TETRANYCHIDÆ.

Gen. TETRANYCHUS.
    Palæarctic Region (Europe).
    Nearctic R. (U.S.A.).
    Neotropical R. (Mexico, Guatemala).

Fam. HYDRACHNIDÆ.

Gen. ATAX.
Palæarctic Region (Europe).
Nearctic R. (U.S.A.).
Neotropical R. (Mexico *, Guatemala).
Æthiopian R. (East Africa †).

Gen. NESÆA (Curriers).
Palæarctic Region (Europe, Bering's Island).
Neotropical R. (Mexico, Guatemala).
Æthiopian R. : East Africa ‡.

Gen. LIMNESIA.
Palæarctic R. (Europe).
Nearctic R. (U.S.A.).
Neotropical R. (Guatemala).
Æthiopian R. (German East Africa ‡).

Fam. BDELLIDÆ.

Gen. BDELLA.
Palæarctic Region (Europe, N.E. Siberia, Bering's Island).
Nearctic R. (U.S.A.).
Neotropical R. (Guatemala, Brazil, Paraguay, La Plata).

Fam. EUPODIDÆ.

Gen. SCYPHIUS.
Palæarctic Region (Europe, N.E. Siberia, Japan).
Neotropical R. (Guatemala).

Fam. IXODIDÆ.

Gen. IXODES.
Palæarctic Region (Europe, Bering's Island).
Nearctic R. (U.S.A.).
Neotropical R. (Central and South America).
Æthiopian R. (South Africa [teste C. L. Koch]).

Gen. AMBLYOMMA.
Nearctic Region (U.S.A.).
Neotropical R. (Central and South America).
Oriental R. (Java, Philippine Islands, Bintang, Continental India [teste C. L. Koch]).

---

* The Mexican species described by M. Alfred Dugès as *Atax alpestri* belongs, according to a private communication of Herr F. Koenike, to the genus *Nesea*.

† Herr F. Koenike, the well-known specialist in this group, writes me that he possesses specimens of *Atax* collected by Dr. Stuhlmann during his expedition with Emin Pasha.

‡ According to the above-quoted communication of Herr F. Koenike.

Australian R. (New Holland [*teste* C. L. Koch]).
Æthiopian R. (South Africa [*teste* C. L. Koch]).

Gen. ARGAS.

Palæarctic Region (Europe, Egypt, Persia).
Nearctic R. (U.S.A.).
Neotropical R. (Mexico, Guatemala, Paraguay).
Æthiopian R. (Ovampo Land).

Fam. ORIBATIDÆ.

Gen. ORIBATA.

Palæarctic Region (Europe, N.E. Siberia, Algeria).
Nearctic R. (U.S.A.).
Neotropical R. (Central America, Brazil, Paraguay).

Gen. HOPLOPHORA.

Palæarctic Region (Europe).
Nearctic R. (U.S.A.).
Neotropical R. (Guatemala, Brazil).

Fam. NICOLETIELLIDÆ.

Gen. NICOLETIELLA.

Palæarctic Region (Europe).
Neotropical R. (Guatemala).

Fam. GAMASIDÆ.

Gen. UROPODA.

Palæarctic Region (Europe).
Nearctic R. (U.S.A.).
Neotropical R. (Central America, Brazil, Paraguay).

Gen. HOLOSTASPIS.

Palæarctic Region (Europe, Bering's Island).
Neotropical R. (Central America, Brazil, Paraguay, La Plata).

Gen. GAMASUS.

Palæarctic Region (Europe, Japan).
Nearctic R. (U.S.A.).
Neotropical R. (Central America, Brazil, Paraguay, La Plata).

Gen. CELÆNOPSIS.

Palæarctic Region (Europe).
Neotropical R. (British Honduras, Panama, Brazil, Paraguay, La Plata).

Gen. PACHYLÆLAPS.

Palæarctic Region (Europe).
Neotropical R. (Mexico, Brazil, Paraguay).

Gen. MENISTHANEA.
>Oriental Region (Malacca).
>Australian R. (New Guinea, Queensland).
>Neotropical R. (Mexico, Guatemala, Paraguay).
>Æthiopian R. (Gold Coast *).

Gen. DERMANYSSUS.
>Palæarctic Region (Europe).
>Nearctic R. (U.S.A.).
>Neotropical R. (Guatemala).

FAM. SARCOPTIDÆ.

Gen. TYROGLYPHUS.
>Palæarctic Region (Europe).
>Nearctic R. (U.S.A.).
>Neotropical R. (Guatemala).

Gen. MEGNINIA.
>Palæarctic Region (Europe).
>Neotropical R. (Guatemala).

Gen. PTERALICHUS.
>Palæarctic Region (Europe).
>Nearctic R. (U.S.A.).
>Neotropical R. (Guatemala, Brazil, Guiana, Ecuador, Patagonia).
>Oriental R. (Himalaya, India, Java, Philippine Islands, China).
>Australian R. (New Holland, New Guinea, New Zealand, New Caledonia, Tahiti, &c.).
>Æthiopian R. (Gold Coast and other parts of Africa).

Gen. PROCTOPHYLLODES.
>Palæarctic Region (Europe).
>Neotropical R. (Guatemala).

There can be no doubt that the geographical distribution of the various families, genera, and even species of many Acarids will eventually be found to be much larger. Several are already known to extend over more than one of the zoo-geographical regions, i. e.:—

>*Atax crassipes*, O. F. Müll.†—Europe, Guatemala.
>*Adlurda terrarum*, L. ‡—Europe, Guatemala, Paraguay (Rio Apa), Buenos Ayres.
>*Holostaspis marginatus*, Herm.—Europe, Guatemala, Brazil (Matto Grosso), Paraguay (Asuncion, Rio Apa), La Plata (Buenos Ayres).

Undoubtedly the migrations of the Acarids may account, to some extent, for the uniformity of types, especially as regards the wholly or partially parasitical species.

---

* I possess a new and very interesting species belonging to this remarkable genus from Acra (Gold Coast), which I propose shortly to describe elsewhere.

† See page 47.   ‡ See page 43.

In some instances, as in the Dermaleichidæ, the passive migration on birds' wings is no doubt the usual mode of dispersion; and as the regions of Central America form a sort of rendezvous and winter station for many birds of passage which are infested by Dermaleichidæ, contact and interchange of Acarid forms of this particular group are continually taking place. Much less extensive, though yet considerable, is probably the passive migration and the dispersion of those Acarids which, in some stages of their existence, adhere to insects and to bats, and which, therefore, are carried by insects or on bats' wings, such as various Tyroglyphidæ, Gamasidæ, Trombidiidæ, and Hydrachnidæ; the Hydrachnidæ attach themselves to Hemipterous water-insects which at night abandon their ponds and take wing. In the course of many successive generations these Acarids may thus spread over large areas. Amongst the non-parasitic species, the active migration of those of open habits and rapid loco-motion, such as the Trombidiidæ, Actinedidæ, &c., may influence the dispersion of types and the mutual penetration into different faunas. How far the aerial trans-portation by wind and storm, which are such powerful agencies in the migration of winged insects and young spiders, may facilitate the passive migration of Acarids is, as yet, entirely unknown.

But, taken as a whole, the various modes of migration, numerous as they are, only serve to explain the similarity of types in more or less contiguous land-areas, such as North, Central, and South America, and the wide range of single species. The almost universal occurrence, however, of certain genera, such as *Ixodes, Argas, Actineda, Trombidium, Rhyncholophus, Holostaspis, Oribata, Atax*, and the world-wide distribution of the fundamental types of Acarids, must have another and more general cause. This, most probably, is owing to the early dispersion of the primary Acarid types from their centres of origin, and in the comparative persistency of those types, due to a relatively perfect correspondence between the once acquired differentiation of their essential organs and their modes of life. Bearing in mind the fact that the local faunas of two so very different and widely separated regions as Central Europe and Central America possess a comparatively large number of identical generic types and of closely allied species of Acarids, we have perhaps a right to generalise and to presume :—

(1) That this uniformity is, geologically speaking, very ancient, and originated in a comparatively early geological period when the relative positions of the continents, the islands, and the seas were altogether different from what they are now.

(2) That the Acarida long ago arrived at that degree of organic development (progressive or regressive) which was the fittest for their various modes of life, not partaking in the great and rapid changes of generic and specific characters which, in the course of the more recent geological epochs, have more or less affected so many of the higher organised types of the animal kingdom.

The presence of *Megisthanus* in Central America, a well-characterised Gamasid genus which has not yet been found in any of the extratropical regions, is a fact of peculiar interest. In his original paper Signor T. Thorell [*] described three species from Java and two from New Guinea. One of the latter (*M. tentudo*, Thor.) has also been mentioned by Signor G. Canestrini [†] as occurring in Queensland. One species has since been added by Signor A. Berlese from Paraguay [‡], and most probably the "Gamase géant" of A. Dugès [§], from Brazil, belongs to the same genus. I have already had occasion to mention the fact that this remarkable type also occurs in tropical Africa [‖]. It therefore belongs to the tropical regions of not less than four different zoo-geographical areas, viz.: the Oriental (Java), the Australian (New Guinea and Australia), the Neotropical (Central and South America), and the Æthiopian (Gold Coast)—a fact which it would be very difficult to explain by a migratory dispersion of recent origin from one starting-point. It is far more reasonable to regard these now so widely dispersed *Megisthani* as the surviving members of a once, that is in former geological periods, coherent group of Gamasids which have been separated in consequence of the slow but material changes of the earth's surface, principally by the successive breaking down of large masses of the earth's crust and the filling up of the thus formed gulfs by the seas.

The genus *Megisthanus* is by no means the only example of the occurrence of one and the same animal type at different regions which at the present time are separated by large tracts of sea, and which for long periods have not had any direct land communication whatever with one another.

[*] T. Thorell, *Descrizione di alcuni Aracnidi inferiori dell' Arcipelago Malese*, 1882.

[†] G. Canestrini, *Acari nuovi o poco noti*, p. 14 (Atti del R. Istituto Veneto di Scienze, Lettere ed Arti, L. II. ser. vi. 1884).

[‡] A. Berlese, *Acari Austro-Americani*, 1888.

[§] A. Dugès, *Recherches sur l'ordre des Acariens*, 3º Mém., 1837.

[‖] See above, p. x, note.

The remarkable distribution of some of the higher animals, such as the *Prosimiæ* among mammals, the *Ratitæ* among birds, the *Crocodilidæ* among reptiles, is well known, and every student of terrestrial Invertebrata who has paid some attention to the geographical distribution of his favourite group must be acquainted with similar facts. I may be allowed here to mention the Gasteropod genus *Clausilia*, the *Nenia* group of which is now limited to the high valleys and mountain chains of Peru, Ecuador, and Colombia, and, with one species only, to the Island of Puerto Rico, and which has its nearest allies not in the New, but in the Old World, in the *Laminifera* group (*Neniatlanta*, Bgt.), which now lives on the top of La Rhune, a mountain near the coast of the Bay of Biscay, and in the *Garnieria* group, the members of which inhabit the mountainous districts of China, Siam, and Cambodia. On some pieces of bark in the virgin forests of the Pacific slope of Guatemala I discovered a new species of *Diplommatina* *, a Gasteropod type, the autochthonous members of which had previously only been known from India and the neighbouring archipelagos. In the woods near Retalhuleu (N.W. Guatemala) I found a new species of the Myriopod genus *Polyxenus*, the type of which is the well-known *P. lagurus*, L. Another species of this well-defined Chilognath genus has been mentioned by Mr. Humbert from Ceylon, and one species has been described from North America by Say. A not less characteristic Myriopod type, the genus *Siphonophora*, abounds under the bark of the fir trees near the summits of the volcanos Agua and Fuego in Guatemala, whilst a nearly allied species has been found in Madagascar by my friend, Prof. C. Keller. Another species has been described from Ceylon by Mr. Humbert. In the woods of the Volcan de Agua, at an elevation of about 10,000 feet, I met with a species of land-leech belonging to the genus *Cylicobdella*, Grube, which is closely allied to, if not identical with, *C. lumbricoides*, Gr., discovered by Prof. Fritz Müller at Desterro in Brazil.

Similar instances of an almost world-wide distribution might, no doubt, be found among other groups of Invertebrata whose facilities for active or even passive migration are very limited.

Unfortunately our present knowledge of the Acarida is too fragmentary to allow any more definite speculations as to the phylogeny of this group. Whether it is, geologically speaking, as ancient as some other groups of the Arachnida, or

---

* This species has since been described and figured as *Diplommatina stolli* in the 'Biologia Centrali-Americana' (Mollusca, p. 20, Tab. I. figg. 19 a, b), by Prof. E. von Martens.

whether it is of a more recent origin, whether its various families, some of which are not very closely allied to each other, took their origin from one or from several types, are questions we shall probably never be able to answer. Arachnoid Arthropods appear early in the strata of the primary periods, and it is quite possible that Acaroid types were among them, though the delicacy and minuteness of their structure made their preservation highly improbable. It is even possible that a closer examination of those sedimentary layers which are fine enough for the preservation of more delicate organisms, such as some of the tertiary strata (those of Oeningen for instance), may lead to the discovery of the larger and more chitinized forms, such as the Ixodidæ, Gamasidæ, and Oribatidæ. At present only one Acarid species is known from the Tertiary deposits of Oeningen. A larger number of Acarid types have been described from the Baltic amber. The brown coal of Rott and the Green River beds of Wyoming have furnished a few isolated forms.

But though, as yet, any direct proofs of the geological antiquity of the Acarid type beyond the Oligocene are wanting, the above-quoted fact of a most extensive geographical distribution of the principal genera, and the general uniformity and similarity of the European and extra-European local faunas, as far as they are known at present, are highly in favour of a pretertiary origin of the Acarid types.

If we compare the Acarid fauna of those parts of Europe where it has been somewhat carefully studied with that of Central America, we are compelled to say that Central America is comparatively poor as regards the number of species, far more so than we should be inclined to anticipate when we take into consideration its great variety of soil and climate and its general richness in vegetable and animal productions. I willingly admit that the districts to which my personal researches were confined are not very extensive, and perhaps other parts of Guatemala, such as the high valleys of Alta Vera Paz, or the forests of the alpine mountains of the "Altos," or even the richer slopes of the Atlantic coast, may have a more varied Acarid fauna; but as the various collectors in other parts of Central America did not, in so conspicuous a group as the Oribatidæ, meet with any other species than those which I obtained in Guatemala, it seems to me probable that even those parts, when searched more carefully, will not prove very much richer or more varied in Acarid forms, though undoubtedly they may yield some new species which escaped my notice. I am therefore inclined to believe that there really exists a comparative scarcity of Acarid species, at least in Western Guatemala, and that the principal cause of this is to be found in the atmospheric and climatic conditions of that country. For organisms of such delicate construction, the hygrometric equilibrium of which is so easily disturbed, the extreme aridity of the "verano" or dry season, which in Western Guatemala lasts for several months, must prove much more fatal than does even the European winter with its frozen and snow-covered ground. On the other hand, the torrential "aguaceros" of the "invierno" or rainy season probably cause the destruction of numerous non-parasitic soft-bodied and unprotected Acari. It is for these reasons probably that we find the greatest variety of species in those genera which are protected to some extent by their entirely or partially parasitic habits, such as the Ixodidæ, Gamasidæ, Hydrachnidæ, and Dermaleichidæ, or which, like *Trombidium*, are capable of a somewhat rapid locomotion, which enables them to reach, in case of need, a shelter to protect them from being drowned or dried up. The various stages of the tough-skinned "Garrapatas" (ticks) are enabled, however, to withstand alike the heat and dryness of the "verano" and the deluges of the "invierno."

The limited number of species is, in some instances at least, counterbalanced to a certain extent by an abundance of individuals. This is the case in some species which are more resistant, or better protected against the influences of the climate, than the majority of their congeners—for example, with *Trombidium muricola*, *Tetranychus guatemala-novæ*, *Atax alticola*, *Amblyomma mixtum*, *Oribata centro-americana*, *Holostaspis marginatus*, and the various species of Dermaleichidæ.

I regret that I had not sufficient opportunities for observing the vertical distribution of the Central-American Acarida. My ascents of the volcanos Agua and Fuego were both made in the dry season, in the unfavourable months of January and February, when, at night, the temperature on the summits was as low as -2° C., and when, from cold and aridity, Arthropod life was reduced to a few species of Insects, Myriopods, and Spiders, these living under the bark of the scattered fir trees and under stones [*]. But, judging from what I have observed in the Swiss Alps, where I found a small number of such conspicuous types as *Rhyncholophus*, *Erythræus*, *Oribata*, and *Gamasus*

---

[*] When I passed the night in the crater of the Volcan de Agua, 19/20 February, 1881, I found in the morning the water in our jar covered with a crust of ice. Under the bark of a fir tree near the edge of the crater I found the dead body of a small species of *Trombidium* mite, which had evidently been killed by the cold. Notwithstanding, I discovered under a flat stone in the crater itself an acarus' nest, the inhabitants of which were winged. This species has since been described by my friend, Prof. A. Pavesi, as *Leptothorax* ....

reaching the snow-line, and where one of them at least (*Rhyacalophus nivalis*, Heer) even surpasses it *, I am inclined to believe that within the tropics Acarids will be found at as high an elevation as any other Arthropod group.

PROF. OTTO STOLL, M.D.

Küsnacht, near Zurich, January 1893.

* According to the observation of the late Prof. Heer, *Rhyacalophus nivalis* reaches an elevation of 9580 feet (top of the Piz Levnarena).

EXPLANATION OF THE PLATES.

TAB. I.

Fig. 1. *Trombidium muricatum* (pp. 1, 44): 1 *a*, palpus; 1 *b*, mandible; 1 *c*, first tarsus; 1 *d*, second tarsus.

2. *Trombidium trilineatum* (pp. 4, 45): 2 *a*, palpus; 2 *b*, tarsus; 2 *c*, texture of the skin.

3. *Trombidium albicolle* (pp. 6, 45): 3 *a*, palpus.

TAB. II.

Fig. 1. *Trombidium hispidum* (pp. 2, 44): 1 *a*, palpus; 1 *b*, mandible; 1 *c*, first tarsus; 1 *d*, second tarsus.

2. *Trombidium gregaricola* * (pp. 4, 45): 2 *a*, palpus; 2 *c*, tarsus.

3. *Trombidium muricola* (pp. 3, 45): 3 *a*, 3 *b*, varieties of the same.

TAB. III.

Fig. 1. *Trombidium nasutum* (pp. 2, 44): 1 *a*, palpus; 1 *b*, second tarsus; 1 *c*, first tarsus; 1 *d*, anterior margin of sternum; 1 *e*, mandible; 1 *f*, texture of the skin; 1 *g*, skin with hairs.

2. *Bdella spinudida* (pp. 15, 48): 2 *a*, tarsus; 2 *b*, top of the rostrum; 2 *c*, mandible.

3. Larva of *Bdella* (?) sp. † (p. 16): 3 *a*, first tarsus, showing the texture of the skin; 3 *b*, mandible; 3 *c*, palpus; 3 *d*, second tarsus.

TAB. IV.

Fig. 1. *Trombidium quinquemaculatum* (pp. 3, 45): 1 *a*, palpus; 1 *b*, fourth joint of first leg; 1 *c*, first tarsus.

2. *Rhyncholophus erinaceus* (pp. 0, 45): 2 *a*, palpus, with top of labium; 2 *b*, section of leg with hairs; 2 *c*, tarsus.

---

* = *T. muricola*, var.   † Probably belongs to *Rhyncholophus*.

BIOL. CENTR.-AMER., Arachn. Acar., January 1893.   c

TAB. V.

Fig. 1. *Actinrda flavrola* [a] (pp. 7, 45) : 1 *a*, ventral surface.
   2. *Actinrda antiguranis* (pp. 7, 45) : 2 *a*, palpus ; 2 *b*, mandible ; 2 *c*, tarsus (lateral view) ;
      2 *d*, tarsus (dorsal view).
   3. *Actinrda retiafirea* (pp. 7, 45) : 3 *a*, tarsus (lateral view) ; 3 *b*, mandible ; 3 *c*, tarsus
      (ventral view).

TAB. VI.

Fig. 1. *Tetranychus gastrunais-nova* (pp. 8, 46) : 1 *a*, palpi ; 1 *b*, tarsus with acabulatra ;
      1 *c*, aricular mandibles.
   2. *Scyphius maculatus* (pp. 17, 48) : 2 *a*, last joint of palpus ; 2 *b*, mandible ; 2 *c*, tarsus ;
      2 *d*, hypostoma.

TAB. VII.

Fig. 1. *Atax altivola* (pp. 9, 46) : 1 *a*, ventral surface ; 1 *b*, dorsal surface of another specimen
      showing unusual position of eggs ; 1 *c*, fourth joint of hind leg, showing the
      swimming bristles ; 1 *d*, fourth joint of first leg ; 1 *e*, claws ; 1 *f*, palpus ; 1 *g*, genital
      lamina.
   2. *Limnesia gastenaltera* (pp. 13, 46) : 2 *a*, ventral surface ; 2 *b*, palpus ; 2 *c*, genital
      lamina ; 2 *d*, tarsus of hind leg ; 2 *e*, top of the mandible.
   3. *Limnesia pulverea* (pp. 14, 46) : 3 *a*, palpus ; 3 *b*, genital lamina ; 3 *c*, tarsus of
      third leg.

TAB. VIII.

Fig. 1. *Atax septem-maculatus* [†] (pp. 9, 46, 47) : 1 *a*, ventral surface ; 1 *b*, mandible ;
      1 *c*, palpus ; 1 *d*, genital lamina ; 1 *e*, hair of first leg.
   2. *Limnesia leta* (pp. 14, 48) : 2 *a*, part of ventral surface, showing the disposition of the
      epimera and genital lamina ; 2 *b*, palpus ; 2 *c*, mandibles ; 2 *d*, tarsus of hind leg.

TAB. IX.

Fig. 1. *Atax septem-maculatus*, var. *ypsilon* [†] (pp. 10, 47) : 1 *a*, ventral surface ; 1 *b*, genital
      lamina ; 1 *c*, palpus.
   2. *Limnesia longipalpis* (pp. 18, 47) : 2 *a*, part of the ventral surface, showing the dispo-
      sition of the epimera and genital lamina ; 2 *b*, mandible ; 2 *c*, palpus.

---

[a] *A. flavrola*, *A. antiguranis*, and *A. retaliora* = *A. bru-nova*, Licm., var.
[†] Probably nymphal stage of *A. altivola*.

TAB. X.

Fig. 1. *Atax dentipalpis*, ♀ (pp. 10, 47): 1 *a*, ventral surface; 1 *b*, palpus; 1 *c*, mandible; 1 *d*, posterior margin of the abdomen, with the genital laminæ (ventral view).

2. *Nraea guatemalensis*, ♀ (pp. 11, 47): 2 *a*, part of the ventral surface, showing the disposition of the epimera and genital arese; 2 *b*, palpus.

TAB. XI.

Fig. 1. *Nraea guatemalensis*, ♂ (pp. 11, 47): 1 *a*, part of the ventral surface, showing the disposition of the epimera and genital laminæ; 1 *b*, genital laminæ; 1 *c*, palpus; 1 *d*, mandible; 1 *e*, section of the fourth joint of hind leg; 1 *f*, tarsus of first leg.

2. *Nraea mexulus* (pp. 12, 47): 2 *a*, part of the ventral surface, showing the disposition of epimera and genital arese; 2 *b*, genital arese; 2 *c*, palpus.

TAB. XII.

Fig. 1. *Amblyomma mixtum*, ♀ (pp. 19, 49): 1 *a*, the same, full of blood after suction; 1 *b*, genital aperture; 1 *c*, anal aperture; 1 *d*, top of mandible (pseudochela); 1 *e*, stigmatic plate (peritrema); 1 *f*, radula of the area maxillaris, viewed from beneath; 1 *g*, tarsus of hind leg; 1 *h*, palpus; 1 *i*, tarsus of first leg.

2. *Amblyomma mixtum*, ♂ (pp. 19, 49): 2 *a*, genital aperture; 2 *b*, anal aperture.

3. *Amblyomma forcii*, ♀ (pp. 31, 50), tarsus of hind leg: 3 *a*, top of mandible (pseudochela); 3 *b*, palpus of the same. (See also Tab. XIV. fig. 3.)

TAB. XIII.

Fig. 1. *Ixodes imarum*, ♀ (pp. 18, 19): 1 *a*, ventral surface; 1 *b*, dorsal surface of a young individual; 1 *c*, sucking-apparatus, showing the palpi, the radula of the area maxillaris, and the mandibles (pseudochela); 1 *d*, tarsus; 1 *e*, stigmatic plate (peritrema). (See also Tab. XIV. fig. 4.)

TAB. XIV.

Fig. 1. *Amblyomma crassipunctatum*, ♂ (pp. 22, 50): 1 *a*, anal aperture; 1 *b*, genital aperture; 1 *c*, stigmatic plate (peritrema); 1 *d*, palpus; 1 *e*, tarsus of first leg; 1 *f*, tarsus of hind leg; 1 *g*, top of the mandible; 1 *h*, radula of the area maxillaris.

2. *Amblyomma rubrurre*, ♀ (pp. 23, 50): 2 *a*, coxa of hind leg; 2 *b*, tarsus of hind leg; 2 *c*, tarsus of first leg; 2 *d*, radula of the area maxillaris; 2 *e*, stigmatic plate (peritrema); 2 *f*, palpus; 2 *g*, top of the mandible; 2 *h*, genital aperture; 2 *i*, anal aperture.

3. *Amblyomma forcii*, ♀ (pp. 31, 50): 3 *a*, stigmatic plate (peritrema); 3 *b*, first tarsus; 3 *c*, genital aperture; 3 *d*, anal aperture. (See also Tab. XII. figg. 3-3 *b*.)

4. *Ixodes imarum*, ♀ (pp. 18, 49), anal aperture. (See Tab. XIII.)

EXPLANATION OF THE PLATES.

TAB. XV.

Fig. 1. *Oribata centro-americana* (pp. 24, 50) : 1 a, ventral surface ; 1 b, lateral view of the cephalothorax with pteromorphae ; 1 c, mandible ; 1 d, palpus with part of the epistomium ; 1 e, tarsus ; 1 f, eggs.

2. *Oribata rugifrons* (pp. 25, 50) : 2 a, ventral surface ; 2 b, mandible ; 2 c, leg ; 2 d, palpus.

3. Larva of *Oribata* sp. (p. 26) : 3 a, mandible ; 3 b, palpus with part of the epistomium ; 3 c, tarsus ; 3 d, hair.

4. *Hoplophora retaliera* (ventral view) (pp. 27, 50) : 4 a, lateral view ; 4 b, mandible ; 4 c, epistomium, with right palpus of the retracted mandibles ; 4 d, palpus ; 4 e, tarsus ; 4 f, eggs.

TAB. XVI.

Fig. 1. *Nicoletiella neotropica* (pp. 27, 50) : 1 a, palpi ; 1 b, microscopical texture of the epiderm ; 1 c, mandible.

2. *Uropoda echinata* (pp. 28, 51) : 2 a, hypostome and palpi ; 2 b, tarsus ; 2 c, microscopical texture of the epiderm ; 2 d, chela of the mandible ; 2 e, tarsus of first leg.

3. *Uropoda inæquipunctata* (pp. 29, 50) : 3 a, ventral view ; 3 b, chela of the mandible ; 3 c, tarsus ; 3 d, anal orifice.

4. *Celænopsis arqumenta*, ♀ (pp. 33, 52), ventral view : 4 a, chela of the mandible ; 4 b, tarsus of first leg ; 4 c, genital orifice ; 4 d, microscopical texture of the skin. (See also Tab. XIX. figg. 3–3 b.)

TAB. XVII.

Fig. 1. *Uropoda centro-americana* (pp. 30, 51) : 1 a, buccal parts in the ventral view, coxæ of first leg, palpi, corniculi labiales, and labium—the whole protected by the anterior margin of the dorsal plate (the mandibles are wanting) ; 1 b, tarsus of first leg ; 1 c, tarsus of second leg ; 1 d, chela of the mandible ; 1 e, sternal plate ; 1 f, spiraculum.

2. Nymph of *Uropoda centro-americana* (pp. 30, 51) : 2 a, ventral view ; 2 b, tarsus.

3. *Uropoda piriformis* (pp. 31, 51) : 3 a, ventral view, without the legs ; 3 b, tarsus ; 3 c, hypostome and palpi ; 3 d, chela of the mandible.

4. *Uropoda discus* (pp. 29, 51) : 4 a, ventral view of the first pair of legs, palpi, and mandibles ; 4 b, palpus ; 4 c, tarsus of second leg.

TAB. XVIII.

Fig. 1. *Megisthanus gigantedus*, ♀ (pp. 31, 51) : 1 a, ventral view ; 1 b, genital orifice ; 1 c, tarsus of second leg ; 1 d, ventral view of the buccal parts.

2. *Megisthanus gigantedus*, ♂ (pp. 31, 51) : 2 a, ventral view ; 2 b, chela of the mandible (deprived of its appendages by maceration in a solution of caustic potash) ; 2 c, ventral

view of the buccal parts; 2 d, genital orifice; 2 e, microscopical texture of the epiderm; 2 f, adhesive grooves at the end of the sternal plate; 2 g, chela of the mandible with its appendages.

TAB. XIX.

Fig. 1. *Megisthanus armiger*, ♂ (pp. 35, 51); 1 a, ventral view; 1 b, lip and palpi; 1 c, chela of the mandible; 1 d, genital orifice; 1 e, tooth-bearing joints of second leg.

2. *Pachylælaps læve*, var. *mexicanus*, ♂ (pp. 57, 52, 53); 2 a, ventral view; 2 b, chela of the mandible; 2 c, genital orifice; 2 d, ventral view of the buccal parts, lip, corniculi labiales, and palpi; 2 e, leg of the second pair.

3. *Celænopsis araneoides*, ♀ (pp. 55, 53), dorsal view; 3 a, chela of the mandible with its appendages; 3 b, lip with corniculi labiales. (See also Tab. XVI. figs. 4—4 d.)

4. *Celænopsis megisthanoides*, ♀ (pp. 55, 53); ventral view of the buccal parts, with the lip, corniculi labiales, and palpi.

TAB. XX.

Fig. 1. *Celænopsis megisthanoides*, ♀ (pp. 56, 53); 1 a, ventral view; 1 b, chela of the mandible.

2. *Holostaspis marginatus* (pp. 59, 53); 2 a, ventral view; 2 b, lip, corniculi labiales, and palpi; 2 c, chela of the mandible; 2 d, tarsus of second leg.

TAB. XXI.

Fig. 1. *Pterolichus mesostoma*, ♂ (pp. 59, 53); 1 a, anal appendages with the adhesive discs; 1 b, chelæ of the mandibles and palpi.

2. *Pterolichus mesostoma*, ♀ (pp. 59, 53); 2 a, ventral view.

3. *Proctophyllodes stellarum*, ♂ (pp. 42, 53).

4. *Proctophyllodes stellarum*, ♀ (pp. 42, 53); 4 a, leg of the second pair; 4 b, chela of the mandible; 4 c, egg with embryo.

5. *Megninia pteroplaxorum*, ♂ (pp. 40, 53); 5 a, nymph of ♀; 5 b, larva.

BIOLOGIA CENTRALI-AMERICANA.

ZOOLOGIA.

Class ARACHNIDA.

Order ACARIDEA.

Suborder I. ACARINA-TRACHEATA, Kramer.

Fam. TROMBIDIDÆ.

[Kramer, Grundzüge zur Systematik der Milben, in Arch. für Naturg. xliii. p. 226 (1877).]

TROMBIDIUM.

Trombidium, Latreille, Gen. Crust. et Ins. i. pp. 14?, 145 (1806).

1. Trombidium mexicanum, sp. n. (Tab. I. figg. 1–1 d.)

Corpus oblongum, pyriforme, depressum, postice rotundatum, tomentoso-hirsutissimum; tomentum + pilis quadrifidis dentum, colore uniformi coccineo, ex seriebus silente; superficies dorsalis impressionibus transversalibus induta. Pedes breviusculi, unvinso-unrini, tomentosi; subtus (exceptis tarsis) pilis longis plumosis in seriem disposito instructi. Palpi longi, margine corporis anteriorem superantes; pilis tomentosis pinnatis induti; appendicula labiferal laterculata, haud lata, longe pilosa. Mandibulæ falciformes, ungue labriferal acuto armatæ. Oculi geminati, in tomento fere absconditi.

Long. 5–6, lat. max. 4 millim.

Hab. MEXICO, Presidio (*Forrer*). Two examples.

Body oblong, pyriform, depressed from above and below; shoulders protracted between the coxæ of the second and third pairs of legs; posterior part of the body cylindrical, its hind margin rounded; dorsal surface with a few transverse impressions; colour uniform, a bright scarlet; the whole body thickly covered with velvety, quadrifid hairs. Coxæ of the second and third pairs of legs separated by the protracted shoulders. Legs rather short, when compared to the mass of the body, of about equal length, the anterior ones a trifle thinner and longer than the others; slightly compressed, higher than broad, thickly covered with short red hairs which give them a whitish silky lustre; beneath bearing brushes of long, straight, pinnate hairs, which in the fore legs are but imperfectly developed and limited to the first three joints (counting from the coxa); the brushes are wanting from all the tarsi; front tarsi obliquely truncate at the top, their lower surface covered with very short hairs which are arranged into a sort of tactile brush, their claws much smaller than those of the other pairs; the tarsi of the second, third, and fourth pairs bearing long pinnate hairs. Palpi long, extending far beyond the front margin of the body; covered with velvety, pinnate hairs, which on the surface of the appendicula form long fringes. The fusiform mandibles are terminated by a broad, falciform, sharply pointed claw, which is much shorter than the mandible itself.

The description of this beautiful species is drawn from two dried and pinned specimens, which were collected by Mr. Forrer.

2. Trombidium hispidum, sp. n.   (Tab. II. figg. 1–1 d.)

Corpus ovato-triquetrum, margine anteriori in spicem triquetrum protracto; immaculatum, tenuiter-hirsutum, uniformiter coccineum; pilis rarioribus, disseminatis, elongatis, pinnatis, ex tomento vestris inclusum, praecipue in regione humerali et versus margines posteriorum. Pedes inaequales, in paribus 1° et 4° longiores; tarsus primi paris fusiformis, unguibus minimis. Palpi exserti, pilis elongatis imbuti; appendicula pyriformis, latiuscula. Mandibula brevis, ungue elongato, angusto, falciformi, subtiliiculos denticulata, apice obtusula.

Long. 2, lat. 1 millim.

Hab. GUATEMALA, Retalhuleu (*Stoll*).

Body triangular, its anterior margin protracted into a sort of triangular clypeus; colour uniform, scarlet. Body and legs thickly covered with a short velvety pile, out of which spring long, disseminate, pinnate, and slightly curved hairs, which on the shoulders and towards the end of the body are longer and more thickly set than on the remaining parts of the dorsum. The legs are comparatively longer than in *T. maricensum*, and without brushes on the lower surface. The front tarsus is fusiform, its claws considerably smaller than in the others; the latter are truncate at their top, and bear pinnate hairs, each of which is inserted on an elevation of the integument. The palpi bear elongate, pinnate hairs, which on the second and third joints are arranged in a row; the appendicula is broader than in *T. maricensum* and bears much shorter hairs. Mandibles short, with a rather cylindrical body, and a long, narrow, slightly curved claw; the concave edge of the latter is, almost imperceptibly, denticulated.

The description and figures are made from living specimens, which were found running amongst dead leaves in the cacao-plantations near Retalhuleu.

3. Trombidium nasutum, sp. n.   (Tab. III. figg. 1–1 g.)

Corpus fere rhomboideum, depressiuscum, postice rotundatum, latitudinem maximam inter coxas secundi et tertii primum paris exhibens, coccineum, tomentoso-hirsutulum; tomentum e pilis brevibus quadrifidis conflatum; pars cephalothoracis frontalis in spicem cartaceum producta, clypei instar imore palporum obtegens; coxae lateribus corporis laxius; utrisque lateris coxae anteriores inter se valde approximata, ab coxis posterioribus (intra se approximatis) ejusdem lateris valde distantes; dorsum impressionum plura transversalium exhibens. Primum et quartum pedum par secundo et tertio paullulum longiora; tarsus primi paris primum fusiformis, unguibus duobus minimis armatus. Palpi marginem clypei superantes; appendicula lanceolata, setas longiores pinnatas gerens. Mandibula brevis, falci brevi, lata, intus subtilissime denticulata armata.

Long. 1, lat. 0·6 millim.

Hab. GUATEMALA, Retalhuleu (*Stoll*).

Body rhomboidal, broadest between the coxæ of the second and third pairs of legs; scarlet, velvety; the frontal part of the cephalothorax forming a sort of fleshy clypeus, which covers the basis of the palpi; a few transverse furrows run across the dorsal surface of the living animal. The coxæ are inserted on the sides of the body; the first and second pairs, which are grouped near together, widely separated from the posterior coxæ formed by the third and fourth pairs of legs. The first and fourth (?) pairs of legs are longer than the second (?) and third, each thickly covered with short velvety hairs. Front tarsi spindle-shaped, bearing two very small claws on the top; the other tarsi obliquely truncate, their claws almost simple. The palpi bear long pinnate hairs, which on the third and fourth joints are arranged into rows; the appendicula is lancet-shaped and bears on its surface several long setaceous hairs, which, however, are rather shorter than in *T. mexicanum*. The mandibles are short and terminated by a short, broad, falciform claw, the inner edge of which is denticulate.

This species is found amongst dead leaves in the forests round Retalhuleu. The description and figures were made from the living animal.

4. Trombidium quinque-maculatum, sp. n. (Tab. IV. figg. 1–1 c.)

Corpus oblongo-ovoideum, depressum, lævissimum; nigrum, maculis quinque albis dorsalibus: harum anteriores partes, magnæ, triangulares, regionem humeralem utrimque corporis lateris occupantes; reliqua tres maculæ imparæ, series formantes longitudinalem in dorso medio; ex his anterior parvula, rotunda, in dorsi centro sita; huic proxima maculæ impar magna, ovalis, in abdominis medio sita; in abdominis apice maculæ magna transversa impar. Apex frontalis cephalothoracis bisetosus longitudinalem, breves, furrolum imprimæ exhibens. Pedum par primum et quartum longæ, crassiuscula, secundæ et tertio pari multo longiores, flavescentia, ab articulo tertio usque ad mentum nigrescentia. Palpi flavescentes, pilis nigris sparsis induci; appendicula lata, lata in seta. Mandibulæ?

Long. 2, lat. 0·75 millim.

Hab. GUATEMALA, near the city (*Stoll*).

Body oblong, a little depressed, with a longitudinal furrow on each side; velvety from closely-set, short, thick, quadrifid hairs: colour deep black, with five white spots; these spots are arranged as follows—two, lateral, large and triangular, occupying the region between the bases of the second and third coxæ; a small round one nearly in the centre of the dorsal surface between the apices of these; a larger ovoid one, placed between the centre and the end of the dorsal surface; and a large transverse one on the posterior end. The first and fourth pairs of legs are very long, about the length of the body, considerably longer and stouter than the second and third pairs; their colour is ochraceous-yellow, which from joints 3–6 merges into blackish, owing to the short, pinnate, black hairs by which this part is clothed. Palpi ochraceous.

I found one specimen of this richly-coloured species amongst shrubs in a ravine near the city of Guatemala; it was running about in the bright sunshine. Unfortunately, I accidentally lost the mounted preparation of the mandibles before I had drawn and described them.

4   ACARIDEA.
bar

5. Trombidium guayavicola, sp. n. (Tab. II. figg. 2–2 c.)

Corpus oblongum, tomentosum; nigrum, maculis duabus albis; anteriori in medio dorso sita semilunari, impari, posteriori ansili, triangulari. Oculi rubri. Pedes ochraceo-rufescentes; par primum et quartum secundo atque tertio longiora, crassiora; tarsorum unguee valde recurvi, ad basin serratuli. Palporum articuli cylindrici; appendicula brevis, recta, apice rotundata. Mandibulae ?

Long. 1, polibus extensis 4, lat. 0·5 millim.

Hab. GUATEMALA, Retalhuleu (*Stoll*).

Body oblong; velvety-black, with two large white spots—a semilunar one on the dorsum between the coxae of the third pair of legs, and a triangular one at the posterior margin of the body. Eyes red; the palpi and legs of a clear reddish-brown colour. The first and fourth pairs of legs longer and thicker than the second and third pairs; front tarsi spindle-shaped; claws of the tarsi strongly curvate, with some indentations at the basis of their concave edge. Palpi straight, slender, with almost cylindrical joints; the fourth joint with a short claw, and a short, straight appendicula, the latter rounded, very broadly inserted, and bearing several long hairs on the top. The fourth pair of tarsi bear a sort of brush beneath, formed of long, oblique, slightly pinnate hairs: these hairs do not fall off so easily as from the other tarsi, and undoubtedly help the animal in running up the trees.

This species is not rare in the forests near Retalhuleu, where it is to be found running up and down the smooth trunks of the guayava trees (palo volador).

6. Trombidium trilineatum, sp. n. (Tab. I. figg. 2–2 c.)

Corpus oblongum, postice rotundato-truncatum, depressiusculum, coccineum, tomentosum, hirsutie brevi, concurrenti; pseudo-prothorax abbreviatus, culco profundo transverso post oculos ab abdomine separatus; dorsum abdominale hirsutie adhaerenti, per sulcos duos longitudinales, laterales, parallelos, evigentes in partes tres distinctas separatum. Pedes, palpi et mandibulae rufescentes. Palpi elongati; tertio articulo cylindrico; appendicula brevi, hoti lata inserta, apice rotundata.

Long. 1·25, lat. 0·78 millim.

Hab. GUATEMALA, Antigua (*Stoll*).

Body oblong, with almost parallel sides; the dorsal surface divided into a pseudo-prothorax and an abdomen by a deep transverse furrow, from which two lateral longitudinal furrows take their origin; these latter divide the back of the abdomen into three separate, whitish areas. Legs long, slender, reddish-brown. Palpi slender, clothed with long, stiff, dispersed hairs; their third joint long, almost cylindrical; the appendicula short, rounded at its end, and with a few stiff hairs on the top.

This species lives in the hedges and gardens of the valleys of Antigua and Guatemala city. It is commonly found on the leaves of bushes, where it seems to feed on Aphides. When the specimens, by rubbing themselves against the leaves, &c., begin to lose their whitish velvety pile, their colour appears much redder than in the specimen figured.

7. **Trombidium albicolle**, sp. n. (Tab. I. figg. 3, 3 a.)

Corpus oblongum, humeris protractis, apice anali rotundato; coccineum, hirsutulum, maculis atque striis albis indutum; in apice frontali pseudo-prothoracis macula alba; stria transversa lata inter humeros, postice linguam latam mediluum in abdomuinis dorsum emittens; apera anale albea; duo maculae albae latero-marginales ante corporis apicem sitae, parva. Coxae lateribus corporis indicae. Pedes longi, graciles, rufescentes, inter se fere aequales. Palpi graciles, pilis longis rarioribus instructi; tertio articulo elongato, cylindrico; quarti ungue unidentato, appendicula recta, apice rotundata, ad apicem piligera.
Long. 0·75-1 millim.

Hab. GUATEMALA, Antigua (*Stoll*).

Body oblong, rounded behind, the shoulders protruding; pseudo-prothorax triangular: colour scarlet, with white spots formed by white, thickly-set, velvety hairs; these spots are arranged as follows—a round white one on the frontal surface of the pseudo-prothorax, a large T-shaped one on the middle of the dorsal surface behind the eyes, an anal one at the end of the body, and two small round lateral ones near the margins of the posterior part of the abdomen. Legs, palpi, and mandibles of a clear reddish colour. Legs long, slender, the first and fourth pairs a trifle longer than the second and third. Third joint of the palpi long, cylindrical; the fourth joint with a small tooth on the concave side of its claw, the appendicula straight, parallel-sided, rounded at the end, where some stiff hairs are inserted.

This species is found with the preceding, on bushes in hedges and gardens in the valley of Antigua.

8. **Trombidium muricola**, sp. n. (Tab. II. figg. 3–3 b.)

Corpus breve, obovoideum, convexum, humeris rotundatis, apice frontali rotundato; cuticula laevis, sericea, nigra, maculis et striis albis, variantibus figura. Pedes, palpi et mandibulae rufescentes. Pedes longi, graciles; primum par coeteris longius.

Hab. GUATEMALA, Antigua (*Stoll*).

Body short, ovoid, convex, with the shoulders and the frontal and anal apex rounded; smooth, silky-black, with white spots and stripes varying in form in the different specimens: in some the dorsal surface of the abdomen bears a white triangular spot on its posterior third, and from the acute angle of this, which is directed forwards, two narrow stripes proceed obliquely towards the humeri; in others there only remains a small white spot at the anal end, and a narrow white streak running across the dorsum at a short distance behind the middle, the streak in its centre forming a large angle (sometimes connected with the anal spot by a narrow white line) which opens towards the frontal apex; finally, there occur specimens in which the whole body is black, except a small white border along the posterior margin. The legs are long and slender, the front pair a little longer than the others. Legs, palpi, and mandibles light reddish-brown.

This species is found in the rainy season on the adobe-walls of the nopal-gardens (small plantations of *Opuntia*, upon which the cochineal insects are reared) round the city of Antigua. It runs busily about in the open sunshine.

RHYNCHOLOPHUS.

Rhyncholophus, Dugès, Rech. s. l'ordre d. Acar. en gén. et le fam. des Trombh. en part., Prem. Mémoire, in Ann. des Sciences nat., Zool. i. p. 15 (1834).

1. Rhynchalophus erinaceus, sp. n. (Tab. IV. figg. 2–2 d.)

Corpus oblongum, convexum ; humeris rotundatis, paullulum prominentibus ; cuticula molli, non refulgente, ex fusco cinerea, pilis elevatabilis brevibus, quadrifidis, nigris band sparse induta ; subtus macula alba prope anum. Palpi, mandibulæ atque pedes rufescentes, piligeri. Palpi articulo extremo longo, curvo ; appendicula longa, lanceolata atque apice rotundata. Epimera primi et secundi pedum paria utrimque lateris in medio corpore valde approximata, per labium fere contigua, ab epimeris posteriorum parium valde distantia ; margo corporis anterior desuper inspicienti globulum piligerum præbens.
Long. 3, lat. 1·5 millim.

Hab. GUATEMALA, Antigua (*Stoll*).

Body oblong, not depressed, with somewhat prominent shoulders, the skin not shining, greyish-brown ; with short, black, quadrifid, thick and slightly curved hairs, which are not very thickly set, so that the skin can be easily seen between them ; the under surface bearing a whitish spot in the anal region. The palpi, mandibles, and legs are reddish-brown, and bear short, black, appressed, obtuse, quadrifid hairs on their upper surface, and longer, imperceptibly pinnate, acute setæ beneath. The fourth joint of the palpi forms a curved tooth, and bears long setæ like the preceding joints and the appendicula ; the latter is obtusely lanceolate in form. When looked at from above, the anterior margin of the body appears to be prolonged into a reddish globe, the latter bearing some long black hairs. The apex of the labium forms a sort of flat cup with many marginal fringes.

Two specimens, both of which were found under stones in the neighbourhood of Antigua.

Fam. ACTINEDIDÆ.

Acaridæ trachealia corpore subrelaxatiori angulis rotundatis, lævi, minima longiore quam latiore, integro. Oculi duo laterales, a margine corporis anteriore valde distantes, facillime perspicui. Cuticula mollis, setis rarioribus stratis, subtilissime pinnatis, induta. Pedes laterales, articulorum nexuum, epimeris pedum in quoque latere valde approximatis. Pedes inter se fere æquales longitudinem, ortis raris vervia et hirsutis brevi, sylæa, appresum induti. Articulos pedum tarsalis gracilescens, acute lævulinæa, unguibus duobus curvis armatos ; ad sorum basin unguis tertius spurius setulosus insertus est. Palpi quaternarum articulorum ; articulus basalis brevis, secundus longus, quam cæteri crassior, subcylindricus, tertius brevissimus, ad apicem interne dentibus trinis, rectis, obtusis armatus, ultimus quam antecedens brevior atque angustior, apice rotundatus, setiger. Mandibulæ longæ, ex basi lata sensim apicem versus gracilescentes, in apice singulo lobiformi armatæ. Inter mandibularum basin in utroque latere corporis scutiter organum longum tubuliforme, angustum, in apice elevatum, quod horizontaliter palpi basin transgrediens marginem frontalem juxta palporum basin ita superat, et a desuper inspicienti destinato et facillime discernatur. Epistoma triangulare, erectum, spice bifido.

ACTINEDA.

Actineda, Koch, Uebersicht d. Arachn. Syst. 3ᵗᵉ Heft, 3ᵗᵉ Abth. p. 57 (1837).

1. Actineda flaveola, sp. n. (Tab. V. figg. 1–1 δ.)

Corpus rotundo-ovoideum, supra convexum, band longius quam latius ; colore citreo, intestinio albido trans-lucentibus.
Long. 0·5 millim.

Hab. GUATEMALA, Antigua (*Stoll*).

Body globose, the dorsal surface convex ; vivid yellow in colour, with some indistinct whitish spots in the middle of the dorsum, caused by the intestinal contents shining through the semi-transparent skin.

This species occurs on the hills round Antigua, amongst the grass.

2. Actineda antiguensis, sp. n. (Tab. V. figg. 2–2 c.)

Corpus breve, subtriangulare vel subpentagonum, paululum depressum, minime longius quam latius ; carmineum, maculis albidis indistinctis ex intestinio translucentibus ; in dorso sulci tres transversi breves tum subis-tantive perspicui ; corporis marginse laterales antice convergentes, ante umbos angulatim flectentes ad marginem frontalem formandum ; margo posterior late truncatus, angulis rotundatis ; hirsutie alba corporis et pedum ita ac in specie praecedenti disposita. Tarsorum unguiculis stylo elongato insertae, curva, integra, ad earum basin utrimque seta pinnata oblique inserta ; ungula tertia species apicem in carunculam infundibuliformem dilatatae.
Long. 0·75 millim.

Hab. GUATEMALA, near the city, Antigua (*Stoll*).

Body short, obtusely pentagonal, slightly depressed from above, a little longer than broad ; carmine-red, with a few indistinct whitish spots on the dorsum : on the latter are three short, transverse furrows, the anterior of which is situated somewhat behind the eyes, the two hinder ones very near each other in the posterior third of the abdomen ; the anterior part from the eyes forward triangular ; the front margin rounded ; the side margins not parallel, slowly diverging towards the hind one, which is broadly truncate, with rounded angles ; on the dorsum are several rows of implanate white shining hairs. Legs covered with short, thinly set, appressed hairs, amongst which many long, stiff, aquartose setæ are conspicuous ; the claws are inserted on a sort of petiole, and bear on each side at their base a fringed short seta ; false claw forming a cup-like caruncle.

This species is found on shrubs in the vicinity of Antigua and the city of Guatemala ; it has a habit of running rapidly up and down the branches of small trees.

3. Actineda retalteca, sp. n. (Tab. V. figg. 3–3 c.)

Corpus breve, subtrigonum, antice rotundatum, postice truncatum, tagulis lateribus obtusis, paululum depressum ; colore rufa, maculis dorsualibus breviroris ab intestinio translucentibus ; in medio dorso inter ocalos latere-marginales macula trigona ; post ocalum utrimque latere stria brunnea longa, antice bifurcata, oblique ad marginem tamen tendens. Tarsorum pulinibus unguiferus medialis tribus dorsualibus.
Long. 0·75–1 millim.

Hab. GUATEMALA, Retalhuleu (*Stoll*).

Body, legs, palpi, and epistoma reddish-yellow. Body short, indistinctly triangular; anterior margin rounded, the posterior one truncate but with rounded angles; on the middle of the dorsum between the eyes is a triangular brown spot, and behind each eye another long and narrow one which goes obliquely to the poster or margin; these three spots are caused by the intestinal contents shining through the soft and semi-transparent skin. The petiole which bears the claws has three nodules on its back.

This species lives in the woods of the low country about Retalhuleu.

Fam. TETRANYCHIDÆ, Kramer.

TETRANYCHUS.

Tetranychus, Dufour, Annales des Sciences nat. xxv. p. 276 (1832).

1. Tetranychus guatemala-novæ, sp. n. (Tab. VI. figg. 1–1 c.)

Corpus oblongum, antice late rotundatum, postice utrolis attenuatum; humeris retundatis, haud protrudis; cuti latero-anteriorum; cuticula tenuis, semipellucida, albidissimo dense pilosula. Palpi crassusi, crassi, trinerum articulorum. Mandibularum in unam minervarum unguiculo in setas longas antice convergentes transformatis sunt, basi reservis. Tarsi unguibus duobus, inter ambulacra quatuor sitis.

Long. 0·75 millim.

Hab. GUATEMALA, near the city (*Stoll*).

Body oblong, its anterior margin broadly rounded; shoulders not or very little protruding; skin semi-transparent, whitish, very finely wrinkled, with some long, regularly disposed setæ amongst many short ones. Palpi short, rather thick, three-jointed, the last joint with a sort of short double claw. The claws of the mandibles transformed into two thin setæ which converge anteriorly. Labium short, bifurcate. Tarsi with two claws amongst four setæ ('ambulacra' of authors), which bear on the top a small, globose bulb.

This species lives in the vicinity of the city of Guatemala, on a common shrub of the genus *Cassia*; it covers the lower surface of the leaves with its silky webs; the yellowish, comparatively large eggs are protected under circular, transparent covers.

N.B.—I regret that I am unable to offer the reader a more exact and complete description of this species, but I discovered it when preparing to leave Guatemala. I had no time left then for the further study of this interesting species, and I would not have reproduced here the above hasty notice and figures, were it not to prove the existence of the genus *Tetranychus* in Guatemala.

Fam. HYDRACHNIDÆ.

[C. J. Nourman, Kongl. Sv. Vet.-Ak. Handl. xvii. no. 8, pp. 16 et seqq. (1880) (Hydrachnidæ).]

ATAX.

Atax (J. C. Fabricius), Nourman, Kongl. Sv. Vet.-Ak. Handl. xvii. no. 8, p. 20.

1. Atax alticola, sp. n. (Tab. VII. figg. 1–1 ♀.)

♀. Corpus obovatum, satis altum, vix depressum, antice et postice rotundatum, postice tuberculis duobus exiguis prominentibus, albo-flavescens, cuticula transparente; macula dorsualis magna, nigra, margine lobata, valde distincta, per glandulam dorsualem bicruriatam in maculas quinque decempra in spinionia divisa; glandula dorsualis antice brunnea, albo marginata, postice flavescens, striam latam medinea longitudinalem formans, ex qua antice et postice rami bini laterales emanantur, iis ut glandula bicruriata apparuat; rami antici latiores, postici angustiores, ramuli irregulares lobiformes emittuntur. Pedes et palpi longi, pedum par primum ceteris paulio crassius; palporum articulus extremus tridentatus. Lamina genitalis arberralum, laten; stigmacibus circiter triancle instructa.

Long. 1 millim.; lat. 0·75 millim.

Mas later.

Hab. GUATEMALA, near the city *(Stoll)*.

Body ovoid, very little depressed, transparent, whitish-yellow; the middle of the back occupied by a large black patch which is divided by the dorsal gland into five distinct spots; the dorsal gland forming a broad longitudinal stripe, brown with whitish margins in its anterior half, yellowish behind; this stripe emitting two lateral branches from its anterior third and two also from its posterior third, the anterior of which are comparatively broad and offer several ramifications, whilst the posterior branches are narrow and bear but a few ramifications. Legs and palpi long, slender, transparent, light greenish. The front legs, which are only a little thicker than the rest, bear a few pairs of long stiff spines (these spines being obliquely serrate towards the apex), and their tarsal joint is furnished with a row of short, acute spines; the second pair has only a few short hairs on the dorsal surface, the lower surface showing several pairs of long spines like those of the first pair; the third pair has on the lower surface numerous pairs of squarrose spines and a short pinnate bristle at the apical end of the fifth joint, whilst the apex of the fourth joint bears a tuft of long swimming-hairs; the fourth pair shows on the lower surface of the fourth, fifth, and sixth joints a row of short broad spines, like the teeth of a comb, and the apices of the joints bear, except in the sixth, a more prominent pinnate bristle and tufts of long swimming-hairs. The palpi bear a few stiff spines; and the obtuse top of their fifth joint is tridentate.

This species lives in ponds in the vicinity of the city of Guatemala.

2. Atax septem-maculatus, sp. n. (Tab. VIII. figg. 1–1 ♂.)

Corpus ovatum, convexum, pellucidum excepta macula dorsualibus, cum pedibus et palpis ex brunneo albescens; macula dorsualis nigro-fusca, magna, per glandulam dorsualem in maculas septem disjuncta;

glandula dorsalis rubra, antice lata, quadrifurcata, postice resorpla duos segmentos laterales emittens. Pedum par primum cubris basal eracitos, par ultimum eximia longina. Laminae genitales parvo, ex imteam cupola briarals longos les, obtuse pentagona; stigmatibus majoribus quiols, minoribus binis inctrosta. Long. 0.4 millim.

Hab. GUATEMALA, near the city (*Stoll*).

Body ovoid, convex, transparent; brownish-white, like the legs and palpi; the dorsal patch large, blackish-brown, separated into seven single spots by the orange-coloured dorsal gland; the latter divided in its anterior half into two bifurcate branches, whilst at its posterior end it emits two small and narrow lateral branches. The first and second pairs of legs bear on their lower surface a few pairs of long, squarrose spines, which are indistinctly serrate at the end and inserted behind tooth-like elevations of the epidermis; the third pair with a pinnate bristle at the apical end of the fifth joint, and also with several swimming-hairs beneath; the fourth pair with a few short spines beneath, the spines gradually becoming longer towards the apex of the joints, and the apical end of the third, fourth, and fifth joints with a short, pinnate bristle, and towards the apices of the same joints some tufts of long swimming-hairs. The genital plates are small, situated in the middle between the epimera of the fourth pair of legs and the abdominal margin, and touch each other at their inner angles; each plate bearing five large and two small stigmata.

Found in ponds in the vicinity of the city of Guatemala.

Var. ypsilon. (Tab. IX. figg. 1-1 a.)

Corpus ovatum, pellucidum; macula dorsali tanquam nigro-fusca, grataloso; glandula dorsalis rubra, Y-formis, integra. Laminae genitales pentagonas, parvulae, angulis interiis se tavicem tangentes, stigmatibus quinis vel senis instructa. Pedus atque palpi sicut in typo. Long. 0.25 millim.

Hab. GUATEMALA, near the city (*Stoll*).

Body ovoid, transparent, with a large brown dorsal patch, which is divided by an orange-coloured Y-shaped dorsal gland into three distinct spots. Genital plates pentagonal, touching each other at their anterior angles, each bearing five or six large and two small stigmata. Legs and palpi as in the type.

Found in the same pond with the type in the vicinity of the city of Guatemala.

I cannot consider this little *Atax* specifically distinct from *A. septem-maculatus*, with which it agrees in the colouring, in the shape of the legs and palpi, and in the character and size of the genital laminae. It differs from typical *A. septem-maculatus* in being a trifle smaller, and in the form of the sharply marked, non-ramified dorsal gland.

3. Atax dentipalpis, sp. n. (Tab. X. figg. 1-1 d.)

Corpus ovatum, leve rotundum, depressulum, pellucidum, flavo-albo-cens, postice tuberculis duobus setigeris; macula dorsalis nigra, lata, par glandulam dorsalem brunneam, albo-emarginatam, longam, antice bifurcatum divisa. Pedus longitudinal, setigeri, secundo et quarto pari caturio longioribus. Palpi elongati, apice tridentato, obtuso; articulo quarto longo, cylindrico, recta dente longo valide, crasso in medio, armato. Lamina genitalis disjuncta, lata, arcuato; stigmatibus senis, in binos acervos ternarum stigmatum segregatis bancrecta.

Long. corp. 0·75 millim.; long. pedum 1·5 millim.

Hab. GUATEMALA, near the city (*Stoll*).

Body ovoid, almost globoid, a little depressed, transparent, brownish-white, with two setigerous tubercles at its posterior end and (in the male) two small tufts of hairs at the end of the genital fissure; dorsal patch black; dorsal gland Y-shaped, brown, with white margins, dividing the dorsal patch into an anterior triangular, and two long, broad, lateral spots. Legs very long and slender, the second and fourth pairs longer than the first and third, all bearing numerous long hairs and rows of short spines. Palpi very long; fourth joint cylindrical, bearing on the middle of its inner surface a strong, obliquely inserted tooth which points forward; fifth joint long, slightly arcuate, with an obtuse tridentate apex. Genital plates situated near the abdominal margin, arcuate; each bearing six large stigmata, which are divided into two equal groups. The epimera of the third and fourth pairs of legs form an almost quadrate plate.

Lives in stagnant ponds in the vicinity of the city of Guatemala.

NESÆA [*].

Nesæa (C. L. Koch), Neumann, Kongl. Sv. Vet.-Ak. Handl. xvii. no. 8, p. 89.

1. Nesæa guatemalensis, sp. n. (Tab. X. figg. 2–2 δ, ♀; and Tab. XL. figg. 1–1 f, δ.)

♀. Corpus ovatum, convexum, pellucidum, albido-flavescens; macula dorsali fusco-nigrescente, per glandulam dorsualem angustam, antice bipartitam, betariam in quinque maculae disjunctam; tres anteriores minores, brevea, duae posteriores majores, margine anteriore arcuatae. Pedes mediocris longitudinalis, a primo pari gradatim longitudine crescentes, sagraturalis valde recurvis; pedum articuli quarti atque quinti fasciculis setarum in seriem obliquam dispositarum ad apicem instructi; pedes posteriores spinarum serie bevrium subtus armati. Palporum articulus secundus crassus, quartus dente armatus. Ovali duo magni, conici, divergentes. Area genitalis stigmatibus senorum, circa 16 magnis atque 5 parvis instructa, quae areae uniformarum in utroque latere fissurae genitalis comparant.

Long. 0·75 millim.

δ. Corpus ovatum, altum, antice obsoleta truncatum; macula dorsali fusco-nigrescente in dorsi dimidio posteriore sita, breviore quam in femina. Mandibulae tertio apicem serrulatae. Pedum unguiculae valde recurvae, ham incrassato incurvo; pedum quartorum articulus quartus in medio excavatus, arrabus duabus dentibus obliqua apposita in marginibus excavationis armatus. Area genitalis stigmata diffuso disposita, circiter viginti.

Long. 0·6 millim.

Hab. GUATEMALA, near the city (*Stoll*).

♀. Body ovoid, convex, whitish-yellow, transparent; dorsal patch brownish-black; dorsal gland narrow, yellowish, anteriorly bifurcate, and dividing the dorsal patch into

[*] This genus requires a new name; it is preoccupied in Polypi (Lamouroux, 1816) and Mollusca (Risso, 1826).

five distinct spots—three short ones before, and two large, longitudinal, parallel ones behind the lateral branches. Legs of moderate size, the first pair the shortest, the following pairs gradually longer; all bearing tufts of swimming-hairs and spines of varied length, and the fourth and fifth joints at their apical ends with tufts of long swimming-hairs arranged in oblique rows; the fourth pair with a row of short spines on its lower surface; the claws strongly curved, forming an acute angle. Palpi with the second joint thicker than the rest, the fourth joint bearing on its middle an oblique setigerous tooth. Eyes conical, diverging. Stigmata of the genital area numerous, about twenty-six on each side, arranged on a semilunar space.

♂. Body somewhat truncate at its anterior margin; dorsal patch limited to the posterior half of the back, the anterior half milk-white. Claws strongly curved, both of them inserted on a common swollen basis. Stigmata of the genital area numerous, about twenty in number, not arranged in a semilunar area as in the female, but rather irregularly disseminate on an obliquely transverse, oblong space. The fourth joint of the hind legs bears on the middle of its inner side an excavation, the margins of which are beset with rows of obliquely set spines or teeth placed opposite to each other. The concave edge of the mandibles shows some very small indentations towards the apex.

The male and female above described were found in a pond in the vicinity of the city of Guatemala. Not having seen them *in copulâ*, nor bred them, I am not quite sure that they belong to one and the same species; yet from their general appearance, their common habitat, and from the fact that I found no other *Nesæa*, except *N. ———*, in this pond, I am inclined to think that the above-described specimens really belong to the same species.

2. Nesæa numulus, sp. n. (Tab. XI. figg. 2-2 c.)

♀. Corpus orbiculatum, admodum valde depressum, antice truncatulum, in margine anteriore utrâque tuberculis trinis piligeris juxta oculam instructum; colore fusco-ovato, opacum, oculis cribelli instar perforatis; maculâ dorsualis latâ, margine exteriore lobatâ, nigrâ, in dimidio anteriore maculis clarâ patrio rotundis interruptâ, in medio per glandulam dorsualem, latam, fuscam, postice utrâque dilatatum divisa. Pedes breves, a primo pari gradatim longiusculos crescentes, setigeri, piligeri; par quartum serie spinarum atque serie pinnatæ ad articuli quarti et quinti apicem subtus instructum; articulos extremum ad apicem subtus subexcavatum. Palpi margine corporis satetiorum paullulum superantes, teres; articulo quarto serrato, dentigero. Stigmata area genitalis numerosa, similes triginta in quoque latere. Oculi in margine anteriore siti.

Long. et lat. 0·75 millim.

Hab. GUATEMALA, near the city (*Stoll*).

Body orbicular, very much depressed, truncate at its anterior margin, the latter bearing three small tubercles on each side (under the microscope the skin seems to be perforated by densely set little holes, like a sieve); colour brown; dorsal patch black, broad, with several transparent brown spots on its anterior half, its exterior margins lobate; dorsal gland compact, broad, brown, truncate before, laterally extended behind. Legs short, the first pair the shortest, the following pairs gradually longer; all bearing a number of spines of varied length and numerous swimming-hairs, the latter being longest in the fourth pair; hind pair with the fourth and fifth joints bearing a row of erect short spines (the most apical of which are pinnate) beneath, and the lower surface of the tarsal joint slightly excavated at the apex. Stigmata numerous, about thirty on each side, disseminate on both sides of the genital plate.

This species lives with the preceding in the vicinity of the city of Guatemala.

LIMNESIA.

Limnesia (C. L. Koch), Neuman, Kongl. Sv. Vet.-Ak. Handl. xvii. no. 8, p. 97.

1. Limnesia guatemalteca, sp. n.　(Tab. VII. figg. 2–2 e.)

Corpus ovatum, altum, pellucidum, late fuscescens, obscure punctulatum maculis dorsalibus quinque fuscis: tribus anterioribus, duobus (aliquando tribus) posterioribus, glandulæ dorsualis albescentis ramulis divisis. Oculi valde distantes, bini in quoque latere inter se vicini. Palpi et pedes fuscescentes. Pedum par primum cæteris crassius, parte trima anteriore articulis extremis obliquæ truncatis, unguiculatis; quarto paro angusturalia serrato, tarsis sparse setigeris, spinigeris. Laminæ genitales parvæ, ab epimeris posticis trigonis distantes; marginæ interiore recto se levissima tangentes, exteriore sinuato; binis stigmatibus in utraque lamina longitudinaliter dispositis. Mandibulæ ungue falsiformi integro. Palparum articulus extremus spice tridentatis.

Long. 0·6 millim.

Hab. GUATEMALA, near the city (*Stoll*).

Body ovoid, high, transparent, light brown, with a fine dark punctulation; dorsal patch divided into five brown spots—three before, two behind—amongst which the whitish dorsal gland spreads its ramifications, the limits of the latter not being so distinct as in many other species. First pair of legs shorter and a little thicker than the others; the tarsal joint of the three anterior pairs obliquely truncate at the apical end. Genital plates small, placed at a little distance behind the posterior epimera, and touching each other along the entire length of their interior margin, their side margins slightly sinuate and diverging; each lamina bearing two large stigmata only. Mandibles with a falciform, narrow, non-denticulated claw.

This species lives in ponds near the city of Guatemala.

2. Limnesia longipalpis, sp. n.　(Tab. IX. figg. 2–2 e.)

Corpus ovatum, pellucidum, album; maculæ dorsuali nigra, triloba; glandula dorsuali subphræna—antice bifurcata, ramis lateralibus valde distinctis, angustis, lobulatis; postice lata, minus distincta. Pedum longitudinis mediocris, a primo pari gradatim longitudine crescentes; tria paria anteriora sparse setigera, articulis tarsalibus obliquæ truncatis; ultimum par ubtus seris setarum minstriciam longarum instructum. Palpi valde elongati; secundo articulo brevi, crasso, dente brevi interno instructo; quarto articulo longissimo, dentigero; ultimo curvato, obtuso, tridentato. Laminæ genitales angustæ, ab epimeris posticis distantes; marginæ interno se levissima tangentes; marginibus externis divergentibus; trinis stigmatibus, uno antico aliis, duobus posticis vicinis, instructæ.

Long. 0·6 millim.

Hab. GUATEMALA, near the city (*Stoll*).

Body ovoid, transparent, high, whitish-yellow, with a black, trilobate dorsal patch; dorsal gland sulphur-yellow—anteriorly bifurcate, and with narrow branches and sharply marked outlines; posteriorly broad, obfuscated, and with rather indistinct outlines. Legs of moderate size, gradually longer from the first to the fourth pairs; tarsal joint of the first three pairs bearing several long hairs and spines, and obliquely truncate; the fourth and fifth joints of the hind legs with a row of long swimming-bristles beneath. Palpi exceedingly long; the second joint rather stout, and with an erect, short tooth; the fourth joint very long, slightly curved; the fifth joint with an obtuse, tridentate top. Internal edge of the falx of the mandibles very finely denticulated. Genital plates narrow, placed at a little distance behind the hinder epimera, and touching each other along the entire length of their interior margin; their exterior margins diverging; three large stigmata on each lamina.

Lives in ponds in the vicinity of the capital of Guatemala.

3. Limnesia puteorum, sp. n. (Tab. VII. figg. 3–3 c.)

Corpus ovatum, altum, opacum, fuscum; maculis dorsualibus nigris, irregularibus; glandula dorsuali compacta—antice alba, longe bifurcata; postice brunnea, indistincta. Pedes mediocris longitudinis, ungulatis ante apicem dente instructis; antici sparse setigeri, postici (parte tertii et quarti) longis setis natatriciibus instructi. Palpi longi, articulo secundo unidentato. Mandibula falci integra. Laminae genitales oblongae, margine interno recto, inter se valde approximatae; stigmatibus tribus magnis instructae.

Long. 1·0 millim.

Hab. GUATEMALA, Antigua *(Stoll).*

Body ovoid, high, not transparent, brown, with irregular black spots and stripes on its back; dorsal gland white and very distinct, bifurcate in its anterior half, brown and indistinct behind. Legs of moderate size, the hind pair with rows of swimming-hairs; tarsal claws bearing a tooth on their concave edge. Palpi long, the second joint with a tooth on its inner side. Claw of the mandibles not denticulated. Genital plates oblong, narrow, touching each other along the entire length of the interior margin; each bearing three large stigmata.

Lives in the " pilas " (water-cisterns) of the city of Antigua.

4. Limnesia læta, sp. n. (Tab. VIII. figg. 2–2 d.)

Corpus ovatum, fere pictistum, altum, pellucidum; maculis nigris dorsualibus trinis in utroque latere—binis laterale magnis, singula externa parva; media in dorso atra lata longitudinalis lata; adest, antice, postice atque in medio transversa alginensa. Pedes haud longi, a prima pari gradatim longitudine crescentes, setis atque spinis variis instructi; par quartum subtus setis spinarum seostatum brevium atque setis natatricibus longis indutum; in apice articuli tertii et quarti ejusdem parte setis pinnatis adest brevis. Palpi breves, articulo quarto dente setigero ornatas obliqua. Mandibula ungulatis falciformibus, angustis, integris. Laminae genitales in cuam, pulygonum, antice femorum brevem lanceolatum profundum contervum; trinis stigmatibus instructas, opimario posticis valde approximatae.

Long. 0·5 millim.

Hab. GUATEMALA, near the city *(Stoll).*

Body ovoid, almost globoid, high, transparent; yellowish-white, with six black spots on the back, these being widely separated by a broad longitudinal yellow stripe which at its frontal and anal end and again in its middle is itself crossed by a transverse blackish stripe, the spots placed thus—two large ones on the anterior half of the dorsum and two others on the posterior half, the latter separated from the former by a white interspace, and on each side at a little distance from the anterior ones is a small one nearer to the margin. Legs of moderate size, rather short, gradually longer from the first to the fourth pair; all bearing many spines and swimming-bristles, which are most numerous on the lower surface of the fourth pair; on the latter beneath there is also a row of short, erect spines on the third, fourth, and fifth joints, together with tufts of long swimming-hairs. Palpi rather short, their fourth joint bearing an oblique tooth. Mandibles with a narrow, falciform, non-denticulated claw. Genital plates united into a single large pentagonal piece, which only in its anterior half bears a lanceolate fissure; three large stigmata present on each side of the genital piece.

Lives in ponds in the vicinity of the city of Guatemala.

Fam. BDELLIDÆ.

[Kramer, Grundzüge zur Systematik der Milben, in Archiv für Naturg. xliii. p. 244 (1877),]

BDELLA.

Bdella, Latreille, Gen. Crust. et Insect. i. p. 153 (1806).

1. Bdella splendida, sp. n. (Tab. III. figg. 2–2 c.)

Corpus oblongum, leve; parte antica magna, trigona, in rostrum longum acutum porrectum desinente; humeri rotundati, post eos corpus penello angustius, usque ad apicem analem gradatim attenuatum; postice late rotundatum; colore ex fusco rubro. Rostrum ad basin paululum inflatum, ad apicem subtus piligerum, setula dimidio paullulum separata, divergens. Palpi quam rostrum longiores, setis sparsis brevibus armati; ad apicem quarti articuli valde elongati setas duas inæquales ferentes, internam breviorem, externam longiorem. Extrema palporum articulus secundo brevior, tertius atque quartus breviusculi. Mandibulæ elongatæ, ad basin latiores, apicem versus gradatim graciliores, thelis minimis instructæ; chelarum dens flexus curvus, acutus; dens mobilis latus, apice truncatus. Pars antica corporis (pseudo-cephalothorax), pedes et palpi coccinei; pars postica (abdomen) ex fusco rubra, striis latis roseis ut lineis tribus transversis angustis inconspicuisque ejusdem coloris induta. Cuticula mollis, subtilissime striatula, in dorso setis sparsis in series longitudinales dispositas gerens.

Long. 1–25 millim.

Hab. GUATEMALA, near the city (*Stoll*).

Body oblong, smooth, gradually narrower from the rounded shoulders towards the anal end, the latter broadly rounded; pseudo-cephalothorax triangular, ending in a long, acute rostrum. The pseudo-cephalothorax, rostrum, palpi, and legs are vivid scarlet; the rest of the dorsal surface from behind the eyes to the anus reddish-brown with darker spots, the latter varying with and depending on the intestinal contents; a broad longitudinal stripe of a rosy light-red occupies the middle of the dorsum, and three unequally placed transverse lines of the same colour run across the back. Palpi longer than the rostrum; the second joint the longest, the third and fourth joints very short; the fifth joint shorter than the second, obliquely truncate at the apical end, where it bears two long stiff setæ, the inner seta being shorter than the outer one. Mandibles very long, broad at their base, narrowing gradually towards the end; chelæ very small, their immovable tooth acute, falciform, the movable one broad, truncate at the end. Skin soft, showing under the microscope very fine and densely-set wrinkles. The dorsum bears several stiff setæ arranged in two longitudinal rows. The point of the epistoma has a small brush of short hairs beneath, which latter, by a narrow median interstice, are divided into two. The hairs of the palpi and dorsum are only most finely fringed, those on the legs being quadrifid. Claws of the tarsi broad, curved; the false (third) claw with a short-haired brush.

This pretty *Bdella* lives amongst dead leaves in the hedges and gardens of the city of Guatemala.

———————————

Note.—On the 10th of July, 1880, I found the larva of an Acarid adhering to one of the fore legs of a large Elateroid beetle, *Chalcolepidius*, sp., in the woods near Retalhuleu, and some days later a stripped-off skin of the same kind of larva on the bark of a tree.

This larva, which I have figured under the doubtful name of *Bdella*, sp. (Tab. III. figg. 3–3 *d*), is 0·5 millim. long, and reddish-yellow in colour, with an ovoid abdomen, which is attenuated towards the rostrum (thus forming a sort of collum, on which a stout rostrum is inserted); the shoulders a little prominent, rounded, and with a large black eye-like spot on each side near the margin; the dorsal surface of the soft abdomen beset with several transverse rows of short, quadrifid, somewhat clavate setæ, and the skin densely and finely wrinkled. The palpi four-jointed: the basal joint short, the second thick, the third narrow, cylindrical, the fourth bearing a falciform claw and a straight appendicula (the latter resembling that of the true *Trombidia*); and with a few pinnate hairs spread over the surface, more numerous hairs adorning the appendicula. Mandibles long, consisting of a large, broad, basal piece, which is attenuated rather suddenly into a long, narrow branch, the latter bearing on its top an extremely small tooth. Legs long, slender, with two curved claws and a pinnate false claw.

Prof. G. Canestrini and Prof. F. Fanzago give in their excellent treatise "Intorno agli Acari Italiani" (Atti Soc. Pad. v. 1877) the figure (tab. 4. fig. 1) and description (pp. 70 *et sqq.*) of an Acarid larva, which bears a strong resemblance to the above-described larva from Retalhuleu, from which it differs, however, in the want of eyes and of a false claw. The learned authors are of opinion that their larva is that of *Rhyncholophus elutoralis*, Koch, or of an allied form. The species from Retalhuleu has in common with *Rhyncholophus* only the short quadripinnate hairs of the back and of the fourth joint of the palpi; it differs from the adult *RhynchoLophi* by the general shape of the body, the insertion of the palpi, the configuration of the mandibles and tarsi, and the finely wrinkled skin.

Fam. EUPODIDÆ.

[Kramer, Ueber Milben, Zeitschr. ges. Naturw. liv. p. 448 (1881).]

SCYPHIUS *.

Scyphius, C. L. Koch, Uebers. d. Arachnidensyst. Heft iii. Abth. 3, p. 62 (1837); Kramer, l. c. p. 449 (1881).

1. Scyphius maniacus, sp. n. (Tab. VI. figg. 2–3 d.)

Corpus oblongum, antice acutum, postice rotundatum, supra convexulum, albidum, semipellucidum, læve, cutis raris in pedibus et abdomine indutum; humeri late rotundati, post eos abdomen contractum. Pedum primum par cæteris longius, par secundum brevius. Pedes corpore margini laterali inserti. Pseudocephalo-thorax triangularis, ab abdomine per sulcum transversum separatam. Oculi nulli. Palpi longi, eorum articulus extremus prosilitur longior, cylindricus, bini marticulus, apice rotundatus, setis pluribus rectis instillans armatus. Mandibulæ epistomium superantes, chelifere, chelarum articulus fixus curvus, dente acuto armatus, ad basin supra setalam gerens; articulus mobilis falciformis, acutus, angustus, haud solum eique extremo apice articulatus fixum tangens. Pedum articuli extremi longi, versus apicem graciliores, unguiculis curvis, unguiculo tertio sparsio, recta, acutam armati.

long. 0·75 millim.

Hab. GUATEMALA, Retalhuleu (*Stoll*).

Body oblong, rather acute towards the frontal end, rounded behind, convex, soft-skinned, semitransparent, whitish, smooth, with a few rows of stiff, minutely pinnate hairs on its back. Abdomen a little attenuated behind the rounded and not very prominent shoulders. Legs inserted on the sides of the body, the first pair longer, the second pair shorter, than the rest; epimera of pairs 1 and 2 separated by an interspace from those of pairs 3 and 4; joints cylindrical, except the sixth, the latter gradually attenuated towards the apical end, and bearing two curved claws and an erect brush-like false claw. Legs and palpi with several squarrose bristles. Palpi long, their last joint rounded at the end, and bearing several long, stiff setæ. Mandibles short, but visible from above; the fixed branch of the chelæ ending in a short curved tooth, which bears a short seta of about the same length at its base outside; the movable branch strongly curved, falciform, narrow, and touching the tooth of the immovable branch only with its acute point; none of the branches bear any denticulations. Eyes wanting.

This species lives in the forests of the "tierra caliente" near Retalhuleu among the fallen leaves. It runs in a very quick but interrupted erratic manner, and is rather difficult to capture. The softness of its skin renders microscopical examination somewhat difficult.

* Præoccupied in Pisces (Risso, 1826).

Fam. IXODIDÆ.

[C. L. Koch, System. Uebers. über die Ordn. der Zecken, in Arch. f. Naturg. x. Jahrg. Bd. 1, p. 220 (1844).]

IXODES

Ixodes, C. L. Koch, loc. cit. p. 221.

1. **Ixodes boarum,** sp. n. (Tab. XIII. figg. 1–1e; and Tab. XIV. fig. 4, ♀.)

♂ latet.

♀. Corpus ovatum, depressum, glabrum, in superficie dorsali et ventrali abdominis tam sulcis longitudinalibus quam punctis impressis instructum ; colore, animalculo vivente, ex cæruleo nigrum (in animalculo alcohole præservato sordide erhraceo). Lamina stigmatica obtuse triangularis, stigmate ex ovali rima, rimam simplicem oblongam formante. Scutum occipitale latum, parvum, angulis obtusis, nitidum, punctis et foveolis raris impressis indutum, colore castaneo uniformi. Oculi nulli. Scutum frontale parvum, glabrum, foveis frontalibus carens, colore castaneo. Rostrum parvum, breve, colore lato castaneo. Arm mandibularis hamulis senis in utroque latere armata, quarto latissimo, quinto longissimo. Arm maxillaris dentibus squamiformibus, rotundatis instructa. Palpi breves, rostrum longitudine band superantes. Pedes graciles, breves, setulis raris armati.

Long. animalous satietatem variaus : 2·6 millim. in jejunis, 5 millim. in satietis animalculis ; lat. 3·6 millim.

Hab. GUATEMALA, Retalhuleu (*Stoll*).

Male unknown. Body of the female ovoid, depressed, hairless. The dorsal surface of the abdomen shows several linear furrows, arranged in two or three concentrical series which radiate from an imaginary centre in the occipital plate. The whole skin of the abdomen perforated by a multitude of extremely small punctiform dimples. Stigmatic plates very simply built, triangular, with rounded angles, the stigma forming a simple excentrical fissure. The occipital plate is small, comparatively broad, with obtuse angles, shining, of a uniform chestnut-brown colour, with some punctiform dimples, principally on the sides. No eyes visible. The frontal plate small, chestnut-brown, devoid of the two frontal grooves or foveæ common to the females of *Amblyomma*. Rostrum small, short. Mandibles bearing six hooks on each side, only five of which are distinctly visible ; the fourth is the broadest, the fifth the longest of all. The maxillary teeth are scaly, rounded. The palpi are rather short ; the legs comparatively short and slender. The total length of the body varies according to the state of digestion ; from 2·6 millim. in the empty specimens up to 5 millim. in the satiated ones. Breadth 3·6 millim.

I found about sixty individuals of this species, all of them females, adhering to the skin of a *Boa imperator*, near Retalhuleu. The rudimentary state of the genital orifice renders it possible that the specimens are not adult.

AMBLYOMMA.

Amblyomma, C. L. Koch, System. Uebers. über die Ordn. der Zecken, in Arch. f. Naturg. s. Jahrg. Bd. 1, p. 223 (1844).

1. Amblyomma mixtum. (Tab. XII. figg. 1–1 i, ♀ : 2–2 b, d .)

Amblyomma mixtum, C. L. Koch, loc. cit. p. 227; Uebers. d. Arachnidensyst. Heft iv. pp. 74, 75, t. 13. figg. 47 (d), 48 (♀).

♂. Corpus ovatum, valde depressum, subtus testaceum, colore ex rufo brunneo. Superficies dorsualis punctis numerosis impressis atque lineolis arcuatis irregularibus, interruptis, undulatis, flavo-albescentibus ornata, quae ex dorsi centro, radiorum instar, ad peripheriam corporis patent. Earum duae maximae angulum immortalem cujusque lateris expromunt inter quas area triangularis fere umbrator et lineolis albidis fere caruns interval. Ferre frontalem nullae. In abdominis parte posteriore lineolae obscure fumae et interstitiis transluceantibus notae, albidis et admixeoral. Inter calculum marginale posterioris lineolae albidae et fumae, brevies, radiatim per paros alternantes disponitae, adsunt. Scutum frontale, rostrum et palpi rufo-brunnei. Orificium genitale rimam acguteam transversalem formans. Laminae analis in utraque latere marginis posterioris seta dualms longis armatae. Pedes rufo-brunnei, setalis punicis seriaris armati.

Long. 4·5 millim.; lat. 3·5 millim.

♀. Corpus ovatum, in jejunis valde depressum, in satiatis depresso-globosum. Scutum coxipitale magnum, triangulare angulo rotundato, centrum dorsi angulo postieo fere attingens, lu-te, colore rufo-brunneo, marginibus lateralibus fusvis, maculis albido-flavis irregularibus, disjunctis, e margine postieo in rumm duo ramificatim disordentibus. Foveolae punctiforme seate impressae transverso, nigra. Abdomen colore fusco-olivaceum, margine clarius brunatero, haud nitidem, setulis albidis rariis ornatum, punctis, setalis et foveis impressis inequum. In jejunio intestino colore fumo translucerat, in satiatis abdomen colorem uniformem obscure purpureum seu rubidam exhibet. Lamina stigmatica triangularis, angulio late rotundato, lateribus arcuatis. Rima stigmatica clariformis. Arce genitalis triangularis; orificium genitale transversale angustum, margine anteriore nobilissima denticulata. Arca analis rectangulata angulis rotundis, valvulis analibus similianalis; quaeque valvula in apice anteriore setis dualms, in posteriore setis tribus armata. Scutum frontale fovvis dualms rotundis ovatum. Palpi, rostrum et pedes in sexubusm oribus aequaliter constructi. Palpi rostrum haud separantes, compressi, articulo quarto brevi, in apice tertii inserto, colore castyoris, ad opicom servandi et ad basin tertii articuli macula fumo interne ornati. Apex rostri albido-flavus. Scuta rostri et anguli posteriores sccti frontalis fusci. Area maxillaris colore scoutore, dentibus concidoria, sat cloretis, inter se sat distantibus armata. Area mandibularis in utraque latere hamulis quinque armata; unam primus longimins, secundum setat appendiculam breves insertum fervus; tertius, quartus et quintus in rumo segregato dispositi. Pedes rufo-brunnea, setalis punctis armati. Pedes antini longiorre, externi drubirohali, ceterorum partum articulos quintos in apice dentibus dualms armatee. Variant feminae hujus species in singulis individuis colore plus minusve clariore ceatorum acxipitalis et frontalis, nec non rostri et palum. Asendit ad hoc quod maculas albido scuti coxipitalis ramificato extensiors multam variant.

Long. 4·5 millim., lat. 3·5 millim. in jejunio, ad long. 12 millim. at lat. 8 millim. in satiatis.

Hab. Mexico (Xorá); Guatemala, Retulbuleu, Antigua (*Stoll*); Nicaragua, Chontales (*Janson*); Costa Rica, Cache (*Rogers*).

♂. Body oval, much depressed, concave on its lower surface, testaceous, the dorsal surface with numerous punctiform black and testaceous dimples and several irregularly shaped, arcuate, narrow whitish stripes, which radiate from the centre of the dorsum and are sometimes interrupted so as to form a mere series of spots: a triangular area behind the anterior margin of the collar is comparatively free from these spots. On each side-margin and parallel to it there runs a narrow longitudinal stripe, which

D² 2

begins near the white eyes and ends in about the middle of the lateral margin. In the hinder part of the body the intestines shine through the semipellucid skin in the form of blackish spots and stripes. On the interstices between the furrows of the posterior margin there occur alternating pairs of black and white linear stripes. The arrangement of the white colouring is subject to individual variation. Colour and shape of the frontal plate, of the rostrum, palpi, and legs, similar to that of the female, except the two frontal dimples, which only occur in the female and are wanting in the male [*]. The genital plate bears a transverse narrow fissure. The anal plate bears two setæ on each side of its posterior margin.

♀. Body oval, in empty specimens much depressed, in the satiated ones globose. Occipital plate triangular, almost reaching the centre of the dorsum, shining, testaceous, with dark brown side margins. From the posterior angle, which is yellowish-white, there proceeds a ramificated branch of the same colour towards the anterior margin of each side: these branches vary much in the different specimens; in some they are broad and continuous, in others they are narrow and tend to resolve themselves into several spots. The abdomen is dark brown, opaque, lighter at its margin, irregularly dimpled and furrowed, and bears short, thinly set whitish hairs. In transparent light under the microscope the ramifications of the intestines are visible in the form of blackish arcuate stripes. In the specimens which are filled with blood the abdomen assumes during life a uniform dark purple hue. The stigmatic plate in both sexes is triangular, its fissure claviform; the stigma proper presents itself as an arcuate small hole in a dark chitinized lamina. The genital plate is triangular, with a narrow transverse fissure, the anterior margin of which is finely denticulated. The anal valvula of each side shows two setæ on its anterior and three on its posterior end. The front plate bears two round dimples. The eyes are white. The palpi are compressed, similar in colour to the body; they bear at the top of the second and the base of the third joint a small brown spot, and are beset with several short hairs; the fourth joint, which is very small, is inserted centrally at the top of the third one. The rostrum is of a light transparent brown, yellowish at its extremity. The mandibles bear five hooks arranged on two branches, the first bearing the first and second hooks, the second bearing the third, fourth, and fifth hooks; the second hook is very short and forms a sort of small appendage to the first one. The maxillary teeth are crooked, obliquely erect, somewhat distant from each other, amber-yellow. The legs are light brown, whitish at the apex of the joints; the first pair have their last joint irregularly denticulate at its apex, and the other pairs bear two teeth at the apex of the fifth joint.

This species is the most common of all the Ixodidæ of Central America, and generally known by the name of "garrapata," which is a corruption of "agarrapata" (clasping something with the legs). I have never found the male in a parasitic state,

---

[*] They have been drawn by mistake in fig. 2 of Tab. XII.

only free on grass and bushes in the " tierra caliente " and " tierra fria " of Guatemala (Retalhuleu, Guatemala city). The female, which abounds in the woods and savanas on grass and bushes, is occasionally rubbed off by horses, cattle, or dogs, and even by man. It adheres tenaciously to the skin, fixing itself by perforating the cutis with its sucking-apparatus; and remains, when undisturbed, for several days, till filled with blood, and then probably falls off spontaneously by its own weight. If forcibly removed, the sucking-apparatus breaks off and remains in the wound, causing a disagreeable and sometimes painful inflammation for a considerable time, but I never saw any serious consequences result from it. Even in its juvenile state the garrapata is of parasitic habits. The young, which are distinguished by the inhabitants of Guatemala by the name of " mostacilla " (derived from " mostaza," mustard), hang to the grass in clusters of thousands, especially during the dry season; and by their creeping on the bare skin and frequent biting they form one of the greatest plagues to the European traveller, who is sometimes kept awake for hours during the night by them. The males I have heard spoken of as " conchuda." The female has been collected by Mr. Janson in Nicaragua, by Mr. Rogers in Costa Rica, and by myself in many places of the " tierra caliente" and " tierra fria" of Western Guatemala (Retalhuleu, Escuintla, Antigua, Guatemala city).

Remarks. Though I have not seen the types of *A. mixtum*, which Koch describes as from Mexico, I cannot doubt that the above described *Amblyomma* really belongs to that species. As Koch describes and figures both sexes, his must therefore have been a common species; and the above described is the most common of all Ixodidæ in Central America, and probably also in Southern Mexico.

2. Amblyomma forall, sp. n. (Tab. XII. figg. 3–3 b; and Tab. XIV. figg. 3, 8 a-8 d, ♀.)

♂ ·····

♀. Corpus ovato-depressum, abdomine cuticula (in exemplo maltato, in clunbolo promorsato) ex griseo fusco aequalitur ·········

Long. (in extinto) corporis 50 millim.; rostri 1 millim, scuti occipitalis 3·5 millim.; lat. corporis 11 millim, scuti occipitalis 2·25 millim.

Hab. GUATEMALA, Retalhuleu (*Stoll*).

Male unknown. Body of the female ovoid, the skin of the abdomen greyish-brown,
marbled with small dark spots and striæ. Occipital plate triangular with rounded
angles, on its dorsal surface with black, comparatively large dimples. The general
colour of the occipital plate is shining black, with a large amber-yellow spot on its
posterior angle, which continues towards the anterior margin as a broad chestnut-brown
stripe. On the humeral margin of the scutum the punctiform dimples are wanting.
Frontal plate shining, black, with a few punctiform dimples and two oblong frontal
holes, truncate behind, angulate in front. Mandibles black at their base, becoming
yellowish towards the apex, bearing five hooks on each side, which are arranged two
and three on two branches. Palpi shining, dark brown, bearing a few short bristles,
abruptly depressed or almost concave on their inner surface, convex externally, the
fourth joint inserted excentrically on the third one. Legs dark brown, with some
bristles. Stigmatic plate triangular, with rounded angles and a claviform stigmatic
fissure. Stigma proper arcuate, with swollen margins.

I accidentally found one specimen of this species among a lot of *A. mixtum* collected
at Retalhuleu. I dedicate it to my friend Prof. A. Forel, the well-known myrme-
cologist.

3. Amblyomma crassipunctatum, sp. n. (Tab. XIV. figg. I, 1 a–1 k, ♂.)

♂. Corpus ovatum, postice late rotundatum, depressum, cuticula ferrolis punctiformibus ut profundis arcte
lærigatis irregulariter intermixtis instructa. Color obscure rubro-fuscus, maculis albidis særatim
marginem corporis distributis: macula magna rhomboides regionem humeralem utriusque lateris occupat,
maculæ quattuor parvæ in interstitiis 3ᵒ, 5ᵒ, 7ᵒ, et 9ᵒ inter sulculos marginis postici siti sunt. In margine
laterali maculæ duæ vel tres adsunt. Scutum frontale supra planum, nitidum, fuscum, lateribus convexo-
declivibus, sine ferris frontalibus, sulculis obliquis irregularibus nec non punctis subtiliter impressis alique
utile punctis albidis ornata. Rostrum et palpi clarius rufo-fusci. Palporum articulus quartus in tertii
apice centraliter insertus. Area mandibularis in utroque latere hamulis tribus in duos ramos distributis.
Hamulorum apices colore saccinco nitentes. Hamulus secundus primi rami bidentatus. Dextæ maxillæ
simplices, æquilongæ transparentes vitreæ, densæ confertæ, squamiformes, medice erecti. Laminæ stigmatis
parvæ, triangulare-semilunares, sæpæ stigmatica parva, clariformi, rima stigmatis arcuata. Valvulæ
anales utrinque laterio arctis quinque armatæ: tribus in margine interno, duabus in margine postice sitis.
Fissura genitalis transversa. In area polygona sita, labris imminentia. Pedes rufo-fusci, ad apices articulorum albido-annulati, setulis armati.

Long. corp. 7·5 millim., scuti frontalis cum rostro 2 millim.; lat. 5 millim.

♀ latet.

Hab. NICARAGUA, Chontales (*Janson*).

Body of the male oval, depressed, dark reddish-brown, with many deeply impressed,
punctiform dimples, between which are a few callous, smooth, irregularly distributed
spots. A comparatively large rhomboid whitish spot adorns the shoulder-region of each
side, and on the 3rd, 5th, 7th, and 9th interstices between the furrows of the posterior
margin there is a small whitish spot. A few irregular spots of the same colour are
seen on the side margin. The frontal plate is flat, shining, with some oblique furrows
and impressed dimples; a few whitish stripes interrupt the light chestnut-brown

colour of the plate. The rostrum and palpi are light chestnut-brown; the fourth joint of the palpi is centrally inserted. The mandibular hooks, three in number, are of a shining amber-yellow at the apex; the second hook of the first branch is bidentate. The maxillary teeth are transparent, uniformly white, simple, aculy, moderately erect. The stigmatic plate is small, its shape between triangular and obliquely semilunar; the stigmatic fissure short, claviform, the stigma proper small and arcuate. Each of the anal valvulæ bears five bristles. The genital fissure, which is comparatively large, lies in a polygonal cavity and is enclosed by a sort of swollen lip. Legs reddish-brown, whitish at the apex of the joints.

Female unknown.

The figures and description are taken from a dried specimen obtained in Nicaragua by Mr. Janson.

4. Amblyomma sabaneræ, sp. n. (Tab. XIV. figg. 2. 2 a-2 i, ♀.)

♂ latet.

♀. Corpus ovatum, depressum. Scutum occipitale nitidum, nei densis et haud profunde punctalatis, colore ex fusco nigrescente, occurrium marginum hemerrabem utriusque lateris maculæ lineari irregulari albedo, quæ in exemplo dealbato difficulter observatur. Scutum frontale nigro-fuscum, foveis frontalibus profundis, ovatis, actius angulatim protractis. Abdomen haud tetrum, crasse punctatum; superficies dorsalis sulcis longitudinalibus latis latique. Rostrum et palpi fusci, clariore apicem versus. Palparum articulus quartus in tertii spina excavata centraliter insertus. Arcus mandibularis hamulis quinque in ramos duos dispositis: hamulus secundus primi rami bidentatus, hamulus primus secundi rami breviludinum. Rostri maxillares acuti, ex dumo conferti, in medio apicem versus torulo ambiguo burrunati. Lamina stigmativa triangularis, angulis rotundatis, fissura stigmatica irregulariter claviformis, lata; rima stigmatis irregulariter ereate. Valvula analis utriusque lateris setis quinque armata. Apertura genitalis margine posteriore recta, lateri regulis integumenti parietis anterioris apertura genitalis tarsi fissuræ recti observantur. Pedes clariore fusco-rufi, sena dentibus duobus conspicientibus armati.

Long. corp. 7 millim. ; scuti occipitalis 3 millim. ; lat. corp. 5 millim.

Hab. GUATEMALA, Retalhuleu (*Stoll*).

Male unknown. Body of the female oval, depressed. Occipital plate shining, punctate, brownish-black, with an irregular whitish stripe on the shoulder-margin, which almost disappears after desiccation. Frontal plate brownish, the two frontal holes deep and oblong. Abdomen not shining, rather coarsely punctate, with some broad longitudinal furrows on its upper surface. Rostrum and palpi brown, lighter towards the apex; the fourth joint of the palpi centrally inserted on the third one, which has its apex excavated. Five mandibular hooks arranged on two branches; the second hook of the first branch bidentate, the first one of the second branch very short and rudimentary. Maxillary teeth acute, with an amber-yellow longitudinal swelling on the middle, their margin double-bordered. The stigmatic plate is triangular with rounded angles; the stigmatic fissure irregularly claviform, comparatively broad, the stigma proper moderately arcuate. Each anal valvula bears five bristles. The posterior margin of the genital orifice is straight, transverse, its side margin convex, arcuate; a series of short linear swellings between the ripples of the integument of the anterior margin. Legs light brown. The coxæ bear two broad, rounded teeth.

I found two females of this species attached to the throat and tail of a small terrapin, known to the natives by the name of " la Sabanera." The figures and descriptions are taken from specimens preserved in alcohol.

Fam. ORIBATIDÆ.

[Dugès, Rech. sur l'ordre des Acar., 3ᵉ Mém., in Ann. Sci. Nat. sér. 2, i. p. 31, *Oribatei* (1834); Kramer, Grundzüge zur System. der Milben, in Arch. f. Naturg. xliii. p. 215 (1877).]

Subfam. *PTEROGASTERINÆ.*

[Michael, British Oribatidæ, pp. 61 & 202 (1884).]

ORIBATA.

Oribata, Latreille, Hist. nat. gén. et partic. des Crust. et des Ins. vii. p. 400 (1804); Gen. Crust. et Ins. i. p. 148 (1806).

1. Oribata centro-americana, sp. n. (Tab. XV. figg. 1, 1 a-1 f.)

Corpus globosum, supra valde convexum, nigrum, læve, nitidum, sine sculptura microscopica ulla. Dorso-rostra ab abdomine non separatus, antice colao transverso ab rostri brevi separatim. Rostrum rectum, valde declivum, triangulare, antice late angulatum, simplex. Lamellæ breves, angustæ, apice brevi, acuto, cum mediocris longitudinis acumine. Pseudostigmata ab dorsupra inspiciunti vix perspiciuntur. Organa pseudo-stigmatica brevia, subdermia, apice pusillulum incrassato. Tectopedia non conspicua. Pedes robusti, mediocris longitudinis, ungulis armati, cœxis, femoribus atque tibiis pusillis complanatis; tarsus ungulius trifum instructus. Pteromorphæ medicæ, antice protractæ, late rotundatæ, plerumque corpori apparent. Iis et a desupra inspiciunti difficulter videntur, colore rufo-fusco, semi-pellucidæ. Laminæ genitales ab caudibus valde distantes, ils minores, Anum rotundato-pentagonum comparatus. Laminæ anales semilunares. Palpi nihil extraordinarii exhibent. Mandibulæ crassæ, breves, brachia chelarum dentibus quatuor obtusis armata. Maxillæ insignium longitudinalem margini anterius rictum profundum prebuit, quæ partem anteriorum marginis externi dextro antico deoursit.

Long. corp. 0·9 millim.; lat. corp. 0·5 millim.

Hab. BRITISH HONDURAS, R. Hondo, R. Sarstoon, Belize *(Blancaneaux)*; GUATEMALA, Antigua, Guatemala city *(Stoll)*; PANAMA, Volcan de Chiriqui 2500 to 4000 feet *(Champion).*

Body globose, of a short and almost circular shape, broad and rounded behind; colour black, the legs and edges of the pteromorphæ brown; texture polished, shining; without any hairs or minute sculpture on the dorsal surface. Cephalothorax comparatively broad; rostrum simple, broadly pointed; lamellæ small and narrow, short, ending in a short acute point; from the anterior margin of the lamellæ there project two comparatively short bristles on each side. Tectopedia wanting. Pseudostigmata short, difficult to be seen from above, projecting from the angle between the basis of the cephalothorax and the anterior insertion of the pteromorphæ. Pseudostigmatic organs moderately large, slightly increasing in size towards the ends, a little recurved on the notogaster. No interlamellar hairs conspicuous. Coxæ and femora of the two posterior pairs of legs rather flattened. Notogaster not separated from the dorsovertex, entirely hairless. Pteromorphæ middle-sized, flexible, semi-transparent, and light-coloured; projecting obliquely forward, and having rounded anterior ends, when seen from the side (from the dorsal aspect their anterior ends seem rather pointed). Genital plates widely separated from the anal plates, occupying a sort of pentagonal area, whilst the anal area is rather circular. Mandibles bearing four blunt teeth on each branch of the chela. Anterior margin of the maxilla showing a deep longitudinal fissure near the outer edge.

This species seems to be common and widely distributed throughout Central America. I have found specimens of it in Antigua and in Guatemala city, under stones and on the moist walls of the house-wells (pilas). It has also been collected in British Honduras by M. Blancaneaux, and on the Volcan de Chiriqui by Mr. Champion. One of the dried specimens from British Honduras which I dissected contained about twenty oval eggs, of 0·3 millim. length.

2. **Oribata rugifrons**, sp. n. (Tab. XV. figg. 2, 2 a–2 d.)

Corpus ovoideum, supra convexiusculum, antice attenuatum, postice late rotundatum, nigrum, nitidum, undum, supra sculpturam microscopicam e rugulis longitudinalibus subtilissimis confusis structam, quæ in cephalothorace magis conspicua, in notogastro paullulum obsoleta sunt. Cephalothorax ab abdomine sulcato transversali discretus, valde declivus, conicus, antice late angulatus, simplex. Rostri clypeus e dorsovertice non discretus. Lamellæ haud conspicuæ. Pseudostigmata recondita; organa pseudostigmatica mediocria, in animalculis desiccato conditorum, apice incrassata. Setæ interlamellares absunt. Tectopedia haud conspicua. Pedes mediocres, colore fusco, setulis et unguibus tribus armati. Pteromorphæ longæ, sat angustæ, antice angulatim protractæ, rotundatæ, postice acutiuscula, mobiles; corpori appressæ pedes supra obtegunt. Abdominis latera ad rудуloydum pteromorphæ læviter excavata. Sculptura microscopica pteromorphorum e rugulis radiatis et insertione cum fundævertibus confluit. Laminæ anales mediocrum, genitalibus majores. Laminæ genitales ab analibus valde distantes, aream rotundato-pentagonam excipiunt. Mandibulæ crassæ, chelarum brachium forte dentibus 4, brachium mobile dentibus 3 argutum. Palparum articulus secundus incrassatus, quintus in medio dentem unguiformem emittens, curvus geminus. Long. corp. 0·55 millim.; lat. max. 0·4 millim.

Hab. BRITISH HONDURAS, Belize (*Blancaneaux*); GUATEMALA, Retalhuleu (*Stoll*).

Body ovoid, convex, rounded behind, its anterior end somewhat blunt; black, shining, but not polished, without hairs on its dorsal surface, this latter showing a minute sculpture consisting of closely placed, extremely fine, longitudinal wrinkles, which are more distinct on the cephalothorax than on the notogaster. A transverse furrow separates the notogaster from the cephalothorax, which is very simply built, conical, ending in a somewhat blunt point. Hood of the rostrum forming but one piece with the cephalothorax. Lamellæ wanting. Pseudostigmata apparently hidden, invisible from above; pseudostigmatic organs in the dried specimens of moderate size, setiform, increasing

towards the end, slightly recurved on the notogaster. Interlamellar hairs and tectopedia apparently wanting. Legs brown, bearing three claws and some bristles.
Pteromorphae long, rather narrow, slightly coloured, brown, semitransparent, their
anterior end protracted forward so as to nearly reach the top of the rostrum, rounded,
the posterior end somewhat pointed; they are very flexible, and when closed completely hide the legs, the whole body then assuming the shape of a small black seed.
The minute texture of the pteromorphae consists of extremely fine wrinkles, which
radiate from their point of insertion. The sides of the anterior part of the abdomen
are slightly excavated for the reception of the pteromorphae. Anal plates larger than
the genital plates, occupying a somewhat pentagonal area. Genital plates semilunar.
Mandibles short, thick; the fixed branch bears four, the flexible branch five, short
blunt teeth. Second joint of the palpi thick; the fifth joint bearing on its middle a
claw-like, pointed, slightly curved tooth.

Examples of this species were captured by me in the forests near Retalhuleu, where
it lives on dead wood in damp places during the rainy season, in company with *Hoplo-*
phora realisca.

Note.—O. rugifrons being the only true *Oribata* observed by myself in the "tierra
caliente" of Retalhuleu, it seems possible that the nymphal form of this genus
delineated by me on Tab. XV., figg. 3, 3 a–d, belongs to this species. Clusters of
hundreds of individuals of these were found crawling on the underside of a piece of
dead wood in a cacao-plantation near Retalhuleu on the 3rd of July, 1880. They bear
a marked resemblance to those nymphal forms upon which Koch founded his genus
Murcia, especially to his *M. trimaculata* (C. L. Koch, Deutschl. Crust., Myr. und Ins.
Heft iii. p. 136. n. 21).

The following are the notes which I then made on the living animal:—

Body oblong, arched, broadly rounded behind, rather diminishing in size towards the
somewhat blunt anterior end. Cephalothorax distinctly separated from the notogaster
by a deep furrow, attenuated at its base, conical, large, triangular; palpi projecting on
the sides of the apex when seen from above. General colour greyish-white, the dark
intestines shining through the smooth and transparent skin. In the angles of the
abdomen on each side a globose, yellow, shining corpuscle. The dorsal surface bearing
a few long pinnate bristles. On the legs the hairs more numerous, also pinnate, but
shorter. First pair of legs the longest, the first and fourth pairs longer than the
second and third. Tarsi with one claw only. Mandibles short, large, the fixed
branch with four, the flexible branch with three teeth. Maxillae of each side bearing
four differently-sized teeth. The pseudostigmata looking like black eye-like spots.
Two long bristles projecting on each side in front of the pseudostigmata. Long.
0·5 millim.

Subfam. *APTEROGASTERINÆ.*

[Michael, British Oribatidæ, i. p. 64 (1884).]

HOPLOPHORA.

[C. L. Koch, Uebers. d. Arachniden-syst. Heft iii. p. 116 (1842).]

1. **Hoplophora retaltena,** sp. n. (Tab. XV. figg. 4, 4 *a*–4 *f*.)

Corpus ovatum, convexum, nitidum, glabrum, colore clarius rufo-fuscus. Cephalothorax conicus, desitrus, ambitu, antice rotundato-acutus. Notogastri latera post cephalothoracem late excisa. Mandibulæ crassæ, dentibus quaternr in quoque brachio armatæ. Dens secundus in brachio fine extra seriem insertus. Maxillarum margo anterior incisuris minus profundis in dentes tres latos divisus. Pedes breves, tarsus unguibus duobus validis, tertio unguiculato armatus.

Long. corp. 1 millim.

Hab. GUATEMALA, Retalhulen (*Stoll*).

Body ovoid, arched, shining, without any hairs on its dorsal surface, reddish-brown. Cephalothorax conical, broadly angular towards its anterior ends, the side margins somewhat rounded. Sides of the notogaster behind the cephalothorax slightly excavated. Mandibles large, with four teeth on each branch of the chelæ; second tooth out of the row on the blade of the chela. Two longitudinal short fissures divide the maxilla into three broad teeth. Legs short; tarsus bearing three claws, two of which are stronger than the other.

Found during the rainy season on dead wood in damp places in the forests near Retalhulen.

Note.—I reproduce the drawing and description of the tarsus, as I made them when in Retalhulen. But still I do not feel quite sure about the correctness of what I saw eleven years ago, as the tarsus of the European *Hoplophora nitens* bears only one claw. Having no preserved specimens of *H. retaltena* at my disposal at the present time I cannot decide the question.

Fam. NICOLETIELLIDÆ.

NICOLETIELLA.

Nicoletia, G. Canestrini & F. Fanzago, Intorno agli Acari Italiani, p. 62 (1877).
Nicoletiella, R. Canestrini, Osserv. sulla *Nicoletiella cornuta*, p. 6 (1882).

1. **Nicoletiella neotropica,** sp. n. (Tab. XVI. figg. 1, 1 *a*–1 *c*.)

Corpus oblongo-ovatum, convexum, antice attenuatum, margine anteriore bicornuta, colore sanguineo. Superficies dorsalis satis sparsis induta. Dorsi cuticulæ ferruginis microscopicis rotundis vel hexagonis, quæ interstitiis transverse striolatis separatæ sunt, regulariter et dense punctulata. Pedes antici arteris longiores atque crassiores, ceteri graciles; pedes postici erratedis atque tertiis paullulum longiores. Tarsi primi

E 2

et segmenti paris ungrtibus duobus, tertii et quarti paris angrstbus tribus armati. Tuberrula lateralia valde conspicua, oblonga. Mandibulis obsletae, crassae, in margine externo sub insertione brachii fixi chelarum dentibus sex brevibus regularibus ornatae. Chelarum brachium fixum curvum, acutum, dente valido ante apicem armatum; brachium valde curvum, in medio serrulatum.

Long. corp. 0·8 millim.

Hab. GUATEMALA, Retalhuleu (*Stoll*).

Body oblong, convex above, slightly diminishing in size towards the anterior margin, which bears two acute cornicles; blood-red, the legs and cornicles slightly coloured; a few short bristles on the dorsal surface, which shows a minute texture consisting of thickly-set round or hexagonal grooves, separated by transversely striated interstices; on each side between the second and third pairs of legs there projects an oblong tubercle. Anterior pairs of legs longer and thicker than the others, which are rather slender; hind legs a little longer than the second and third pairs. The tarsi of the two anterior pairs of legs bearing two, those of the third and fourth pairs three claws. Mandibles short, thick; the fixed branch of the chelæ slightly curved, pointed, bearing a strong tooth near its apical end; the movable branch strongly curved, its inner edge serrulate from the middle to the base; under the base of the fixed branch the body of the mandible shows a row of six short regular teeth.

Found on one occasion only among fallen leaves in a forest near Retalhuleu.

Fam. GAMASIDÆ.

[P. Mégnin, Monogr. de la famille des Gamasidæs, in Journ. de l'Anat. et de la physiologie (1876).]

Subfam. *UROPODINÆ.*

UROPODA.

Uropoda, Latreille, Gen. Crust. et Ins. I. pp. 157, 158 (1806).

1. Uropoda echinata, sp. n. (Tab. XVI. figg. 2, 2 *a*–2 *e*.)

Corpus supra et infra convexulum, subglobulare, paullolum longius quam latius, dorsi scutum ovale, curvum, colore brunneo, ab corporis margine membrana albido-grisea lata separatum, punctulis densis microscopicis nec non setis brevibus indutum. In membrana conjunctiva marginali series (ex oculorum microscopicis ostendit, in quarum quoque seta brevis, spiniformis sdest. Tarsus pedum paris primi ungulbus curti; anteriorum parium unguss gracilior, acuti. Palpi simplices, breves, crassi. Mandibulae longe protractiles, graciles, chelis ungualis, parvis. Chelarum brachium fixum mobili est longius, ad apicem dilatatum, ambo brachia dentibus obtusis indistinctis ornata. Scutum ventrale convexulum, foveolis punctatum.

Long. 0·75 millim.

Hab. GUATEMALA, Antigua (*Stoll*).

Body almost globular, a little longer than broad; the dorsal and ventral plates chiti-

coat, reddish-brown, widely and completely separated by a whitish membrane, the latter bearing several rows of small conical protrusions, on each of which is a short thorn-like bristle; the dorsal plate also shows several short bristles and a dense microscopical punctuation. The legs are slender, comparatively long; the tarsi of the first pair are without claws, the remaining tarsi with two slender claws. Palpi short, 6-jointed, simple, rather thick. Mandibles, when protracted, long and slender; the fixed branch of the chelæ is longer than the movable one, both branches being indistinctly and bluntly denticulated. Ventral plate rather convex, dotted with small grooves.

One specimen, found crawling among mould under a hedge in the vicinity of Antigua, Guatemala.

2. Uropoda inæquipunctata, sp. n. (Tab. XVI. figg. 3, 3 a–3 d.)

Corpus valde depressum, oblongo-ovatum, antice in apicem brevem protractum, postice late rotundatum, nitidum, colore coccineo. Dorsum microscopio inspicienti foveolas punctiformes sparsas ostendit, inter quas puncta minuta numerosa, marginem versus eoque numerabilia sita sunt. Pedes breves, longitudine fere æquales, tarsi primi paris unguibus carentes. Mandibulæ longæ, graciles; chelæ graciles, parvæ, earum brachia irregulariter atque subtiliter denticulata.

Long. 0·3 millim.

Hab. GUATEMALA, Retalhuleu (Stoll).

Body flat, oblong, its anterior border forming a short acute angle, its posterior border broadly rounded, shining, scarlet-red; the dorsal surface shows several punctiform grooves and the space between them is very finely punctured by numerous microscopic points, which are more numerous near the borders. Legs short, almost equal in length; anterior tarsi without claws, the tarsi of the remaining legs with some thorn-like spines. Mandibles long and slender, with small chelæ, the branches of which are irregularly denticulated.

Found at Retalhuleu as a parasite on a coprophagous beetle (*Pinotus* sp.) called "ronron" by the inhabitants.

3. Uropoda discus, sp. n. (Tab. XVII. figg. 4, 4 a–4 c.)

Corpus planum, fere orbiculare, antice leviter protractum, brunneum, nitidum. Palpi breves, in animalculis cum moventi marginem anteriorem parallelum se porrectæ, eorum articulus quartus tertio longior, ad apicem antice nec non arcis verrucosis, granulatis, circumscriptis æqualiis ornatus. Pedum par primum breve, asterculiforme, unguibus carens, ejus articulus secundus atque tertius dentibus obtanis brevibus irregulariter serratis, articulus ultimus æquatus. Mandibulæ longæ, teneræ, translucido-albidæ; chelæ curvæ-flaviæ, parvæ, subtiliter denticulatæ.

Long. 0·3 millim.

Hab. GUATEMALA, Retalhuleu (Stoll).

Body flat, almost orbicular, its anterior margin slightly protracted, but round. Colour of the body, palpi, and legs brownish-red, shining; mandibles white, transparent, their chelæ yellowish. Palpi rather short, their last joint showing several granulated

spots and some bristles. Mandibles long and slender, the branches of the chelæ apparently
deeply denticulated. Anterior pair of legs rather short, without claws, their second and
third joints irregularly denticulated; the posterior pair of legs longer than the others.
I found one example of this species in a decayed chichique-fruit in the woods near
Retalhuleu in June 1880. It is nearly allied to the European *U. cassidea*, Herm.

4. Uropoda centro-americana, sp. n. (Tab. XVII. figg. 1, 1 a-1 f; 2, 2 a, 2 b, nymph.)

♀ *juv.* Corpus depressum, oblongo-ovoideum, antice latus secundum et tertium pedum par obliquo subrotundatum, ita ut pars anterior anguli dorsualis in angulum obtusum protrahatur. Superficies dorsualis plana, marginem versus declivis, foveolis numerosis sculpta, ex quibus setulae breves enascuntur, colore brunneo claro (in desiccato animalculo), minus nitente. Pedes omnes unguiferi, sat breves. Articulus ultimus primi paris pedum setis numerosis creatis, inter quas una longissima juxta unguem sita, unguem in margine longe gracili inserti, graciles. Articulus ultimus ceterarum parium pedum setulis sparsis atque spinis duabus juxta unguem pediculum armatus, brevis. Palpi breves, setigeri, in articulo penultimo seta longa, bifida. Palpi maxillares bidentati. Sub apice anteriore scuti dorsualis setulae duae insertae. Mandibulae graciles, chelarum brachia fere longitudine aequales, dentibus paucis et obtusis in margine interno undulata.

Long. 1 millim.; lat. 0·6 millim.

Hab. NICARAGUA, Chontales (*Janson*).

Body (of the young female) depressed, oblong, its anterior margin forming a somewhat
obtuse angle; dorsal surface even, marked with numerous grooves, out of each of which
a short bristle arises; the colour, in dry specimens, a clear reddish-brown, not very
shining. Legs rather short, all of them bearing claws; on the tarsi of the first pair
the claws are fixed on a sort of slender petiole and surrounded by numerous long hairs,
one of which is very much longer than the others; the tarsi of the second, third, and
fourth pairs bear short bristles and two incrassated spines at the base of the petioles of
the claws. Palpi short, bearing numerous long bristles on their last joint; on the
inner side of their third joint there is a long, projecting, pinnate hair. Mandibles
slender, their chelæ small, blunt at the top, bearing on the inner edges of the brachia
four or five obtuse teeth.

(N.B.—Adhering to the posterior abdominal segments of a Guatemalan specimen of the
Coleopterous genus *Atractocerus* I found numerous nymphs of a *Uropoda* (figg. 2, 2 a, 2 b)
which I am inclined to identify with the above-described *U. centro-americana*, as they
have most of the characters in common, except some which may be the result of the
differences of the respective stages of development. These differences are as follows :—

Body flat, oblong, not so distinctly angular on its anterior margin, yellowish, transparent, the hairs on the dorsal plate apparently a little longer and less numerous ; the
incrassated spines of the legs (which in the adult are placed at the base of the claw)
attached to the side of the tarsus, one of them on the false joint and the other on the
tarsus proper. Length 0·4, breadth 0·25 millim.

Clusters of these nymphal forms attach themselves by viscous threads to the skin of their host, the threads being strong enough to keep them from falling off, even when dried up or preserved in spirit.

6. Uropoda piriformis, sp. n. (Tab. XVII, figg. 3–3 d.)

Corpus ovato-piriforme, subconvexum, subtiliter reticulatum ... (*in moderate*), pilis brevibus paucis in margine ornatum. Oriticula foveolis numerosis crebrata. Palpi simplices, uniguri, in articulo basali seta crassa armati. Mandibularum chelae dentibus numerosis in brachiis ambabus armatae, Orificium genitale subtriprosideum.

Long. 1·2 millim., lat. max. 0·9 millim.

Hab. MEXICO, Jalapa (*Höge*).

Body longer than broad, somewhat piriform, a little convex on the back, naked, except for a few short scattered hairs on the margin; dorsal and ventral plates rather coarsely punctured. Palpi simple; one strong bristle on the inner side of the basal joint. Anterior margin of the epistome lobed; two strong bristles on the ventral surface of the exterior lobes.

One specimen.

Subfam. GAMASINÆ.

MEGISTHANUS.

Megisthanus, T. Thorell, Descriz. di alc. Aracnidi Inferiori dell' Arcipel. Malese, in Ann. Mus. Genova, xviii. p. 49 (1882).

1. Megisthanus gigantodes, sp. n. (Tab. XVIII. figg. 1–1 d, ♀; 2–3 ♀, ♂.)

♂. Corpus longius quam latius, larvarum subovatum, antice attenuatum, ita ut subpiriforme apparent, supra mollius convexum, nitidum, ...

[illegible] duarum spinarum signalant. Cuticula ad coxarum basin in dentis formam paullulum producta. Femora quarti paris supra tuberculis quatuor cuspidatis [illegible] munita sunt, ex quibus setae fortes spiniformes oriuntur; tibiae apicem versus dentes tres fortes gerunt. Ceteri articuli setis sparsis muniti, ex quibus una ad apicem articuli quarti, altera ad apicem quinti articuli longiores sunt. Epistoma magnum, triangulare, supra transverse compressum, antice paullo deflexum, apice in angulum acutum paullo productum, eandem texturam microscopicam exhibens quam scuta ventralia. Ab animalculo vivo epistoma ad latera [illegible] capitis [illegible] movetur. Palpi longi, teretes, setis sparsis quae in apice sexti articuli magis numerosae sunt, muniti. Hypostoma in utroque latere dentem [illegible] obtusum ostendit, in medio in lingulam longam, angustam, quadrilobatam, [illegible], pellucidam, desinit. Mandibulae fortes, earum chela longa, angusta, brachiis ambobus denticulatis; in brachio fixo circa 12, in mobili circa 14 dentes observavi. Chelae brachium fixum in superficie interna [illegible] quae ex chela media [illegible] videtur; apex brachii mobilis membranula tenuissima, hyalina, convoluta (mappula apud Thorell) ornatur, cujus margo subtilissime serrulatus est. In superficie laterali brachii chelae mobilis [illegible] tres [illegible], quorum duo sub apice brachii siti, et valde inter sese approximati sunt, tertius paullo distans, in medio chelae fere situs. Ad basin externam brachii mobilis lamina ex spiralis hyalinis [illegible] difficillime observatur. Dentes labiales longi, curvi, integri.

Long. corp. cum epistom. 3·5 millim., sine epistom. 3 millim.; lat. max. 2–2·25 millim.

♀. Corpus ovatum, antice truncatum, supra convexulum. Scutum dorsale integrum, [illegible] foveolas impressas, ex quibus setulae breves, [illegible], oblique retrospicientes. Limbo albido ad latera et postice a scutis ventralibus separatur. Color laete brunneus. In utroque latere prope margines scuti dorsalis series [illegible] longarum inserta est. Textura microscopica e cellulis [illegible] constante, qui scutum totum areolis oblongis irregulariter reticulatum, sicut in mare scuta ventralibus. Venter in medio convexus, lateribus in [illegible] regione depresso-concavis. Scutum sternale angustum, antice et postice rectum, lateribus ob [illegible] incisuras sinuatis, a scutis lateralibus interstitio angusto, a scuto anali interstitio lato albicanti discretum. Inter coxas secundi atque tertii pedum paris orificium genitale situm est, magnum, subpentagonum, margine postico [illegible]. Scutum anale margine antico recto, lateribus atque margine postico rotundatis, orificium anale in medio situm. Scuta lateralia [illegible] ab anali interstitio albido lato separata, postice angulatim producta, apice rotundata. Epistoma, palpi et pedes sicut in mare. Hypostoma in medio in lingulam quadrilobatam hyalinam productum, dentibus lateralibus, qui in mare adsunt, carens; cuticula laterali sinuata, quae palporum coxulam a labii trunco separat. Palpi labiales longi, graciles, apice acuto, curvo, margine interno dente uno [illegible]. Mandibulae sicut in mare, atque brachio mobili chelae, qui [illegible] dentium [illegible].

Long. cum epistom. 2·5 millim.; lat. max. 1·25 millim.

Hab. GUATEMALA, Acritunno, Guatemala city, Zapote (*Champion*), Retalhuleu (*Stoll*). In regione "Cholhuitz," hanc speciem in coleoptero-pectinicorni "*Proculo goryi*" denominato parasitam inveni.

♂. Body longer than broad, rounded behind, truncate at its anterior margin, much depressed, smooth and shining; a few hairs are placed near the anterior margin, and a row of six or seven obliquely inserted long bristles adorn the hind margin on each side; colour light reddish-brown, in some individuals darker than in others; the dorsal plate is separated from the ventral pieces by a whitish border on its side- and hind-margins, and shows a minute sculpture consisting of small, irregularly polygonal areas with a few impressed points between them; ventral surface convex in the middle, depressed at the sides, smooth, its border whitish and very narrow. Sternal plate narrow; rather long, smooth; its anterior margin truncate, its hind margin rounded. Genital orifice situated between the coxæ of the 2nd and 3rd pairs of legs, small, circular. The sternal plate has two grooves at the hind margin, one on each side, out of each of which a strong hair arises; the grooves are encircled by an apparently elevated broad whitish border, the border showing a microscopical granulation.

These characters are peculiar to the male sex of the genus *Megisthanus*, and are wanting in the female; the colour of the sternal plate itself is much lighter than that of the anal plate, and it is somewhat irregularly distributed—darker on the anterior and lateral parts, lighter on the middle; the anal orifice is situated in the middle of the anal plate. The side-plates extend beyond the hind coxæ, forming a rounded angle at their end. The anterior legs are antenniform, long, and without claws; at the top of the tarsi there are some densely-set tactile hairs. The second and third pairs of legs are strong, somewhat fusiform, bearing on the back of the femora some spine-like hairs. The posterior femora have on the back an equidistant row of four spine-like hairs and on their lower surface three strong, short, blunt teeth. The remaining joints of the legs are irregularly set with hairs. The epistome forms a large triangular hood, which is convex above and ends in a short acute angle; during life this piece can be moved upwards and laterally; it hides the oral parts but imperfectly, as it is surpassed in length by the palpi and, sometimes, by the upper part of the mandibles. The palpi are long and simple, sparingly set with hairs, which are more numerous at the top of the fourth and fifth joints. The margin of the hypostome bears a blunt tooth on each side, and its middle projects into a transparent quadrilobate tongue-like piece. The mandibles are strong, their chelæ rather narrow, bearing numerous irregular teeth on the inner edge; on the inner surface of the fixed branch of the chela a transparent narrow brush projects a little over the top of the mandible; the movable branch seems to be enveloped at the top in a small, transparent, irregularly-folded membrane, the border of which is serrulated; on the lateral surface of the movable branch there project three transparent appendages, each of which bears some finger-like ramifications, and at the base there is a transparent cluster of small spinules.

♀. Body ovate, its anterior margin truncate, slightly convex on the dorsum; dorsal plate but moderately shining, rather coarsely punctured, a short semi-erect bristle arising from each puncture; colour reddish-brown, the border of the dorsum whitish. On each side of the dorsal plate, along its lateral margin, a row of long bristles is inserted. The microscopical texture is similar to that of the ventral plates and hood of the male; it consists of a network of fine anastomosing furrows, which leave long, narrow, irregular areæ between them. Ventral surface convex in the middle, depressed and concave at the sides. The whitish interstice between the sternal and lateral plates is narrow, that between the sternal and anal plates being broader. The genital orifice, which is comparatively large and pentagonal, lies between the 2nd and 3rd pairs of coxæ. The anterior margin of the anal plate is straight, its side-margins being rounded; the anal orifice is situated in the middle of the plate. The lateral plates project into an obtuse angle behind the hind coxæ. Epistome and palpi as in the male. The hypostome bears no lateral tooth; the tongue-like piece is formed as in the male. The movable branch of the chela has fewer teeth than in the male; but the fixed branch and appendages are similarly formed.

This is the largest Gamasid known as yet from Central America. In the Cholbuitz district near Retalhuleu I found it parasitical upon a very large pectinicorn-beetle, *Proculus goryi*, Guér.; but it cannot be confined to this species only, as the genus *Proculus* is not known to occur in the localities where Mr. Champion obtained his specimens of *M. gigantodes*. It is perhaps parasitic on other Passalidæ.

2. Megisthanus armiger. (Tab. XIX. figg. 1, 1 a–c, ♂.)

Megisthanus (sic) *armiger*, Berlese, Bull. Soc. Ent. Ital. xx. p. 201, t. 9. figg. 1, 1 a (1888).

Hab. Mexico, Jalapa (*Höge*).—Paraguay, Rio Apa.

Among the dried specimens of Gamasidæ received from Messrs. Godman and Salvin there is one male *Megisthanus*, collected by Herr Höge at Jalapa, which I believe to belong to the above-named species, as it coincides with the description in all its essential characters, as well as in the measurements, though it differs from it in some minor details.

The differences are as follows :—

Specimen (♂) from Jalapa (Mexico).	Specimen (♂) from Rio Apa (Paraguay), as described by Mr. Berlese.
Scuti dorsalis superficies pilis brevibus e foveolis protuberantibus non nisi punctulis impressis ornata. In dorsi medio area oblonga triangularis subtilis lineari circumscripta adest.	Scutum dorsuale et margines pilis curtulis vestita. (Nothing more mentioned.)
Ante scutum sternale continbus prosternalis, parvum, transversum, medio longitudinaliter partitum situm est.	(Not mentioned.)
Scutum dorsale postice rotundatum, anterius excurvatum. Scutum ventrale est ab anali (transverso quadrilatero) disoretum.	...... rotundatum, in medio tamen elevatiom, anterius transcriptum. Scutum ventrale ....., parum ab anali (triangulari, vertice latiore) discretum.
Margo cartapediorum internus excurvatus.	(Straight in Berlese's drawing.)
Scuta ventralia omnia setulis brevibus vestita.	Mesogathia et scutum ventrale sctis vestita, cetera nuda.
Femora tertii paris interne calcaribus tribus duobus obsoletis, quarti paris calcaribus quatuor brevibus obtusis aucta.	Femora tertii et quarti paris interne calcaribus tribus validis aucta.
Coxisculi labiales curvi.	Coxisculi labiales recti.
Trochanter secundi paris superne inermia.	Trochanter secundi paris superne spina valida armatum.
Cuticula in articulos quarto et quinto secundi, tertii et quarti pedum paris satis ampla; in dentium squamiformum subtus cuiulerata.	(Not mentioned.)
Articulus palporum secundus dente brevi ante apicem subtus armatus.	(Not mentioned.)

CELÆNOPSIS.

Celænopsis, Berlese, Acari, Myr., et Scorp. Ital. fasc. xxxi. no. 5 (1886).
Diplogynium, Canestrini, Atti Soc. Ven.-Trent. xi. p. 101 (1888).

1. Celænopsis uropodoides, sp. n. (Tab. XVI. figg. 4–4 d, and Tab. XIX. figg. 3, 3 a, b, ♀.)

♂ latet.

♀. Corpus oblongo-ovatum, depressum, supra nitidum convexum, postice indistincte acuminatum. Scutum dorsale limbo angustissimo, submitidum, setis brevibus in superficie et in margine subserratis munitum. Color indicus. Sculptura microscopica e minulis anastomosantibus confecta, qui areolas oblongas irregulares circumscribunt. In harum arcolarum medio parvis seriatim disposita impressa sunt. Palpi et pedes marginem corporis superant. Superficies ventralis depressa, eadem structura microscoporum sicut dorsalis præbens. Scutum ventrale magnum, postice acuminatum, marginem posticum corporis non attingens, orificium anale prope apicem ferens. Coxæ utrinque lateris approximatæ. Inter coxas secundi atque tertii pedum pars orificium genitale (a clarissimo bujus generis auctore "epigynium" appellatam) in parte anteriore scuti ventralis sitam est, magnum, subpentagonum, antice latum, rotundatum postice rotundatum. In epigynio distinguendæ sunt:—(1) crista epigynico jugiformis, lateribus deductis, transversa, antice armata; (2) valva valvæ utrinsque lateris aleformis, interne atque antice rotundata, ad angulum internum anticam et prope marginis externi medium seta munita; (3) scutum parvum trigonum a margine postice epigynii inter valvas utrinsque lateris paullo productum; (4) fulcra valvæ claviformis, obliqua, armata. Ante epigynium scutum sternale heterogonum situm est, cujus margo exterior excavatus, lateralis in angulum rotundatum productus est. Ante angulum et prope marginem posticum in utrosque latere setula inserta est. Scuto laterali segmenta, longa, prope marginem corporis posticum rima triangulari segmenta, albide separata. Spiraculum parvum, prope coxas postimas externe sitam. Pedes primi paris graciles, coram latum unguibus curvus, longus, setis numerosis ad epicrum tectibus, sicut articulus palporum externus, munitus. Pedes ceteri lævimus, cremicros, catalis maxilli. Palpi laevimus, non longi. Epistoma trigonum, apice in mucronem acutum desinens. Mandibularum chelæ brachiis gracilibus, dentimulatis; dentes brachii mobiles inæquales; ad apicem parvuli, confertti, infra inter se distantes, majores, pauci; in brachio fixo subæqualis, apicem versus magis confertti ac brevicem. Hypostoma in linguam antice rotundæ trunutam desinens, et que processum duo longi setiformes nascuntur. Caruliculi labiales inni lata in curns extroraus curvum sutuns polbrichum desimans. Brachium mobile membrana pellucide marginibus siumnda latet impartiris obtectum. Apparatus ribarium totus difficillime cernitur.

Long. 0·75 millim.; lat. 0·60 millim.

Hab. British Honduras, R. Hondo (*Blancaneaux*).

Male unknown. Body of the female ovate, depressed, moderately shining, bearing some short suberect hairs on all parts and showing a microscopic texture, consisting of a network of very fine irregular anastomosing furrows, which leave long irregular spaces between them, each space having a series of impressed points. The ventral plate is large and acute behind, and does not reach the end of the body. Anal orifice small, oblong, situated in the posterior angle of the ventral plate. The space between the two rows of coxæ rather narrow. The genital orifice (called "epigynium" by Mr. Berlese) is situated in the anterior portion of the ventral plate; it is large, somewhat pentagonal in shape, broad at its anterior margin, rounded behind. The sternal plate is hexagonal, excavate at its anterior margin, the lateral margin forming a rounded angle. The lateral pieces are long and narrow, separated from the ventral piece by a

F 2

narrow interspace ; at the end of the body there is a small triangular space between them. The spiracle is small, distinctly visible, and situated outside the hind coxa. The legs are without spines or teeth, and bear only some hairs and bristles, which at the top of the first pair of tarsi (and on the palpi) are long and more thickly set than on the other joints. The epistome is triangular and protrudes from under the anterior margin of the dorsal plate as an acute point; it imperfectly covers the mandibles, the branches of which are irregularly denticulated.

Several specimens of this species, which is closely allied to the South-American *C. subracisa*, Berlese, were collected at Rio Hondo by M. Blancaneaux.

2. Calamopsis megisthanoides, sp. n. (Tab. XIX. fig. 4, and Tab. XX. figs. 1, 1 *a, b,* ♀.)

♂ *latet.*

♀. Corpus ovatum, depressum, sertice et postice declivate, postice subacuminatum, antice, propter epistomatis apicem, in angulum acutum desinens, lateribus in regione boureveli sinuatis. Color clare badius. Scutum dorsuale integrum, antice truncatum, ejus margo in lateribus et in margine posteriore ; limbus angusto albido ab scutis ventralibus separatum, metis sat longis sparsis la superficie armatum. Scutum sternale parvum, transversum, inter coxas primi paris sitam, antice recte truncatum, lateribus et margine postico sinuatis. Scutum ventrale magnum, marginem posticum fere attingens. In ejus parte anteriore vulva, rimam bifidam simulans, sita est. Pedes cuneus antulis sparsis moniti fere inermes, exceptis femoribus quarti parisque calcaribus tribus valde obtusis at seta valida haud approximata armata sunt, nec non tarso secundi paris qui ad spinam juxta unguium pedunculum dentem curvum exhibet. Pedes antici longi, graciles, antici crassiores, in certum femorum dorso seta duo valida minant. Epistoma trigonum, declivum, antice mutatum, ad ejus basin in margine scuti dorsualis seta fortis in utroque latere adest. Corniculi labiales longa, recurvirem, antice recurvi, dentibus curvatis. Inter cornua basin aliquo hypostoma procurrens spadi- lliformis scutum transparum adest. Hypostoma in lingnam longam, angustam, tripartitam desinens. Chelarum brachia longa, acgusto, irregulariter denticulata, brachium mobile dentem sciumum intus propo basin exhibens. Palpi setigeri, inermes.

Long. 1·25 millim. ; lat. 0·75 millim.

Hab. PANAMA, Bugaba (*Champion*).

Male unknown. Body of the female ovate, depressed, somewhat acuminate behind, the lateral margin in the region of the shoulders somewhat sinuated ; anterior border of the dorsal plate truncate, bearing a strong bristle on each side over the basis of the epistome ; the lateral and hind margins of the dorsal plate show a narrow whitish border ; the dorsal surface bears some comparatively long but thinly-set hairs. The sternal piece is small, its anterior margin truncate, the lateral and hind margins sinu- ated ; it occupies the space between the first and second pairs of coxae. The ventral plate is large and almost reaches the hind margin of the body. The vulva is situated in the anterior portion of the ventral piece ; it forms a sort of bifid fissure. The fore legs are long and slender, the other legs shorter and thicker ; all of them are without spines or teeth, except the second pair, which bear a small curved tooth at their top, and the femora of the hind pair, which underneath bear a row of three somewhat blunt teeth ; between the base of the hind femora and the first of these teeth there is a strong bristle. The labial horns are long and slightly curved, in the shape of the letter **g** reversed; between their bases and the hypostome there is another process, the top of which forms an acutely-pointed transparent foliole or spatula; the hypostome ends in a long and narrow tongue-like process.

Two specimens.

PACHYLÆLAPS.

Pachylælaps, Berlese, Acari, Myr., et Scorp. Ital. fasc. li. no. 10 (1888).

1. Pachylælaps hæros.

Pachylælaps hæros, Berlese, Bull. Soc. Ent. Ital. xx. p. 199, t. 8. figg. 2, 2 *a–e*.

Var. mexianus. (Tab. XIX. figg. 2, 2 *a–e*.)

♂. Corpus late ovatum, supra convexulum, antice subangulatum, postice rotundatum, antice et postice declivum. Color crustarum, rostri pedumque badius, membranarum conjunctivarum albicans. Scutum dorsale integrum, fere nudum, dense reticulatum, retis spatiis in parte anteriore et in marginibus minutum, a margine postico membrana conjunctiva sistelas crumarum curvula separatum. Pedes primi paris graciles, articulis plus minusve cylindricis, retis spatiis inæqualia, tarsis unguiferis. Pedes secundi paris crassiusculi, dentibus seu calcaribus ut sequitur armati: dentes breves duo in coxæ apice adsunt, trochanter est inermis, femur calcari fuscu, curto, indistincte tridentato armatus, genu atque tibia tuberculum parvum bigibbosum gerunt, tarsus unis apicem dentem fortem obtusum præbet. Coxæ tertii paris dentibus duobus, uno centrali, altero marginali, brevibus munitæ, trochanterus ad apicem dentes duos juxtapositos dengere, tertium infra ferunt, ceteri articuli pilis simplicibus inermes. Coxæ quarti paris dentibus duobus brevibus sicut in tertio pari munitæ, trochanterus ad apicem duosque dentes duos subcostos juxtapositos, infra dentem unum obtusum gerunt, femur ad basin et in apice dentes armatum, ceteri articuli, tarsis inermibus exceptis, dentibus duobus armati. Scutum sternale cum ventrali concretum reticulatum, postice trimmio-rotundatum, marginibus posticus cum siliqueus, ab anali separatum. Scuta metapodica parva, trigona angulis rotundatis. Scutum anale parvum, trigonum. Orificium genitale parvum in margine anteriore scuti ventralis sistum. Inter scutum ventrale et hypostomo lobulum parvum linguæformis, scutalis duobus ad apicem munitus mobilis insertus est. Chelæ mandibularum mediæ, dentibus paucis armatæ, in brachio mobili processum longus ad chelæ basin in apicem pluries emrulatas insertus est. Corniculi labiales distincte biarticulati, eorum articulus postremus unis apicem dentem parvum gerit. Palpi inermes, excepto articulo primo inferne aliis robustum ferente.

♀. Statura et colore maris; a mare præcipue differt pedum pari secundo coxæ in basi crassiore, inermi, ut et quarto pari calcaribus curvulo, inermi.

Long. 2 millim.; lat. 1·6 millim.

Hab. Mexico, San Andres Tuxtla (*Sallé*).

Body of the male broadly ovate, slightly convex, its anterior border forming a short acute angle; colour of the hard plates, the rostrum, and legs light reddish-brown, the connecting membranes whitish; the dorsal plate almost hairless, bearing only a few bristles on its anterior and lateral borders; the bristles are longer and more thickly set on the whitish membrane which connects the dorsal plate to the ventral pieces; the dorsal plate shows a minute reticulation, consisting of a network of small polygonal areæ; almost on the middle of the dorsal surface there are two slightly impressed grooves. The first pair of legs are slender, their joints almost cylindrical, bearing some hairs, but without spines or teeth. The second pair of legs are very thick, the femur especially, the latter bearing a strong, straight, slightly denticulated spine on the middle of its under surface; the genu and tibia each bear a small tubercle, and the tarsus is armed with a broad laminar tooth at some distance from the tip. The third pair of legs are comparatively slender, without spur, except a short one on the upperside of the trochanter at the tip. The fourth pair of legs are longer than the third, their trochanters bearing two teeth on the upper surface at the apex and a short tubercle below; the femora bear underneath a tooth at their base and one at their apex; the fourth joint has a blunt tooth and a short spine at its tip. The sternal and ventral plates form but one piece of an irregular shape, which protrudes but little behind the fourth pair of coxæ and ends at some distance from the anal plate; this piece has its side-margins excavated for the reception of the coxæ, its posterior portion being widened laterally and truncate behind; the anal piece is small, triangular; the lateral pieces are large and bear the distinctly visible spiracle; the metapodial piece is small, of a triangular or rather oviform shape. The palpi are unarmed, except their first joint, which bears a strong spur on the inner side. The labial horns are distinctly two-jointed, and have a tongue-like process between them. The chelæ bear a few strongly-marked teeth at their edges; from the centre of the movable branch a long process issues, the end of which is rolled up somewhat spirally.

The female differs from the male in the following characters:—All the legs are slender and, with the exception of the apical teeth of the trochanters, do not bear any spurs or teeth; the sternal plate is separated from the ventral piece; there are no metapodial plates; and the dimensions of the body are a trifle smaller.

Two specimens, one male and one female, of this large Gamasid were collected by M. A. Sallé at Tuxtla. Their affinity to *P. horres*, Berlese, from Brazil, is so remarkable that I can only treat them as a variety of that species. The Mexican specimen (*d*) before me differs from Berlese's description in the following characteristics:—

P. horres, typ. *d* (as described by Berlese).	Var. *mexicana*, *d*.
Spur of the femur of the second pair thick, curved, with two teeth at the tip; spur of the genu of the second pair comparatively elevated, distinctly visible; tarsus of the second pair without a tooth at some distance from the apex; genu of the fourth pair bearing but one tooth (not mentioned in the description, but present in the drawing).	Spur of the femur of the second pair straight, denticulate at its margins; spur of the genu of the second pair short, rudimentary, forming only a tubercle; tarsus of the second pair bearing a distinctly visible compressed tooth at some distance from the apex; genu of the fourth pair bearing two teeth—one pointed, spined (like the lateral teeth), and one blunt, on the ventral surface of the apex.

HOLOSTASPIS.

Holostaspis, Kolenati, Wien. ent. Monatschr. ii. p. 67, t. 1. figg. 1, 2 (1858); Canestrini, Prospetto dell' Acarofauna Italiana, p. 55 (1885).

1. Holostaspis marginatus. (Tab. XX. figg. 2, 2 a–e.)

Acarus marginatus, Herman, Mémoire Aptérologique, p. 76, t. 6. fig. 6.

Holostaspis marginatus, var. *americanus*, Berlese, Bull. Soc. Ent. Ital. xx. p. 195.

Hab. GUATEMALA, Retalhuleu (*Stoll*): NICARAGUA, Chontales (*Janson*).—SOUTH AMERICA, Brazil, La Plata, Paraguay.

Two specimens of this classical *Gamasus* were collected by Mr. Janson at Chontales. I have frequently met with it at Retalhuleu, where its various larval and nymphal stages live as parasites on the common "roaron" or dung-beetle (*Pinotus* sp.), together with the nymphs and larvæ of *Gamasus fucorum*, De Geer (*coleoptratorum*, auct.).

Subfam. DERMANYSSINÆ.

PTEROPTUS.

Pteroptus, L. Dufour, Ann. Sciences Nat. xvi. p. 98 (1832).

The bats of Central America are, as well as those of Europe, infested by parasitical mites of the genus *Pteroptus*, which attach themselves to the smooth skin of the wings and the axillary cavity. But as my drawings are too incomplete to allow the identification of the species, I do not reproduce them here, but merely state that this subfamily is represented in Guatemala.

Suborder II. ACARINA-ATRACHEATA, Kramer.

[Kramer, Grundzüge zur Systematik der Milben, in Arch. für Naturg. xliii. p. 318 (1877).]

Fam. SARCOPTIDÆ.

[Michael, British Oribatidæ, p. 50 (1884).]

Subfam. TYROGLYPHINÆ.

[Trouessart, Les Sarcopt. plumicoles, 1ᵉ part. p. 6 (1885).]

In Guatemala I have occasionally observed the presence of free-living *Tyroglyphi*, as well as of hypopial forms attached to the body of several species of flies (Muscidæ), but I have not taken exact notes as to any of them.

Subfam. *ANALGESINÆ*

[Trouessart, *Les Sarcopt. plumicoles*, 1ᵉ part. p. 5 (1885).]

MEGNINIA.

Megninia, Berlese, Acari, Myr., et Scorp. Ital. fasc. iv. (1883).

1. **Megninia pteroglossorum**, sp. n.　(Tab. XXI. figg. 5, 5 *a. b.*)

♂. Corpus ovoideum, minus ex albido-griseo, pellucidum, convexulum. Pedum paris anterioris inter se æquales, breves. Pedes tertii incrassati, longi, abdominis marginem posteriorem superantes, articulo ultimo acuto, unguifero, ceteromula adhærentia indato, ceteris articulis cylindricis. Pedes quarti breves, tenues, in articulo ultimo ungues duo fervotes, abdominis marginem vix attingentes. Appendices anales lamelliformes, acuti, margine interiore integro, exteriore itinerario duabus sed profundis sinuatis. Pars posterior appendicularis utrimque lateris setas duas longas ferens, seta brevior in lobo inter incisuras eminente; longissima juxta basin appendiculæ externæ sita. Disci adhæsivi prope marginem postirum setis appendicularis in ipso abdomine siti, magni. Epimera primi pedum paris in lineæ medianæ ad laminam sternalem formandam sese conjungunt.

Long. corp. circa 0·6 millim.

♀ nympha. Corpore oblongo ovato, antice latiore, marginem posticum versus attenuatum, postice rotundatum. Pedes breves, corpore margini inserti, sparse setigeri. Pedes anteriores crassiorum, breviores, posteriore latiorum, longiorum. Epimera primi paris in lineæ medianæ cum tangentia, secundi paris brevia, libera. Margo posterior abdominis integer, setis quatuor armatus, interius brevioribus, externis longioribus.

Long. corp. circa 0·5 millim.

Larva hexapoda corpore oblongo, margine integro, super pedum insertionem angulariter protracto.

Hab. GUATEMALA, Retalhuleu (*Stoll*). In remigum vexillis *Pteroglossi torquati* parasita gregatim vitam degit.

Body of the male ovoid, greyish-white, transparent, shining, convex. Anterior pairs of legs of equal length, short. Third pair of legs very thick, long, protruding over the hind margin of the body, their last joint pointed, bearing claws, the other joints cylindrical. Fourth pair of legs short, slender, with two teeth on their tarsus, hardly reaching the end of the body. The anal appendices are pointed, lamelliform, their inner margin straight, their outer margin sinuated, with two profound incisures; the posterior portion of the appendices bears two long bristles, and there is a short bristle on the lobe between the two incisures; the adhesive discs are large and situated near the posterior margin of the body before the appendices. The epimera of the fore legs of both sides are united into a common sternal plate.

Body of the female (nymph) oblong, somewhat thinner towards the posterior end, rounded behind. The legs are short and inserted at the margin of the body; the two first pairs are a little thicker and shorter than the hind pairs. The epimera are brown and have the appearance of narrow hard bands; those of the fore legs of each side reach the median line, those of the second pair remain separated. The hind margin of the body bears four bristles; in young specimens it is a little excavated.

The hexapod larva is oblong, the side-margin forming a sort of angle over the insertion of the legs.

This species lives gregariously as a parasite in the wing-feathers of the "Cucharon" (*Pteroglossus torquatus*) in the tierra caliente of Retalhuleu (Guatemala).

PTEROLICHUS.

Pterolichus, Ch. Robin et Mégnin, " Mém. sur les Sarcoptides plumicoles," Journ. de l'Anat. et de la Physiologie, 1877, p. 393.

1. Pterolichus momotorum, sp. n. (Tab. XXI. figg. 1, 1 a, b, d; 2, 2 a, 2.)

♂. Corpus convexum, translucidum, ex albido griseum; margines circumferentia irregulari inter coxas secundi et tertii paris latas retrorsumque corporis ad lobi instar protrusum, inbaln in anteriorem minorem atque posteriorem majorem dimaknam. A basi tertii paris pedum corpus ad basim appendicum analium corpus crassicadim diminuit. Appendices anales magnae, basi lata, margine externo integro, convexa, margine interno irregulari, in medio rotunde ascens, dimula. Appendiculae apice ant obtuse, octis duabus longis nec non spinis tribus brevibus munita. Disci adhaesivi in appendicularum basi siti. Pedum antum inter et sequales, omnibus latum caruncula magna in pedunculo tenui inserta instructus. In dorsi medio super coxas secundi pedum paris acta quaetuor, duo interna brevten, duo externae longae, sita sunt. Spina valida recta in latere corporis ante coxas tertiarum pedum. Palpi biarticulati. Mandibula breves, crassa, carum chela dentibus paucis brevibus armata. In facie ventrali sutum sternum parvum, inter coxas pedum anteriorum, atque aculum ventrale, ab illo interstitio clariore separatum, oblongum, distinguuntur. Epimera anterior, linearia, coloro brunneo. Epimera primi et secundi paris oblique versus lineam mediam tangentia, sed coxa non attingentia, apicibus late separatis. In ventris medio trabecula semicircularis ante orificium genitale sita.

Long. corp. circiter 0·5 millim.

♀. Corpus majus atque longius quam in mari, lateribus ad lobi instar protrusum, post coxas quarti paris cylindricum, lateribus parallelis, apice rotundato, in medio rotunde excisa, justa excisionem in utroque latere setis duae longae et spina breves duae adsunt. Epimera, uta dorsi, palpi, mandibulae atque pedes ut in mari.

Long. corp. circiter 0·5–0·75 millim.

Hab. GUATEMALA, Retalhuleu (*Stoll*), in vexillio remigum *Momotus Lessoni* gregatim.

Body of the male convex, transparent, of a milky-whitish colour; the side-border is irregular and projects between the second and third pairs of legs in the form of a bipartite lobe; from the base of the third pair of legs the abdomen diminishes in size. The anal appendices are large, broad at their bases, convex externally, their inner border irregularly sinuated, with a round excavation in its middle; they are rather blunt at their tip and each bears two long bristles and three short spines; the adhesive discs are situated at their base. On the anterior portion of the back four bristles are inserted, and the side-margin bears, at a short distance from the third pair of coxae, a straight short spine. The legs are of about equal length and shape, each tarsus bearing a large caruncle. The palpi are two-jointed, and the mandibles are short and thick, their chelae armed with a few short teeth. The ventral surface is divided by a clearer interstice into a smaller sternal piece, occupying the space between the anterior coxae, and a larger and oblong ventral piece. The epimera form narrow, pointed stripes of a brownish colour, which are directed from the side-margin towards the median line of the body.

The female differs from the male by its larger size and the more oblong shape of the body, the abdomen being cylindrical from the hind legs, and by the want of the anal appendices. The abdomen is simply rounded at the apex, the hind margin bearing a small round excision in its middle. The legs, the dorsal hairs and lateral spines, the epimera, the palpi, and mandibles are as in the male. A chitinized stripe of brownish colour and semicircular shape is situated transversely before the genital orifice.

This species lives gregariously in the wing-feathers of the " Pajaro bobo " (*Momotus lessoni*). In running, the female takes the lead and draws the smaller male after her. In the act of copulation the female places her abdomen above the anal appendices of the male. (In figg. 1, 2 this connection already appears loosened a little.) I observed this species in November 1880.

P. momotorum bears, by the form of its anal appendices, a remarkable affinity to *P. phylloproctus*, Trouess. et Mégn., which inhabits the *Haliëtus leucogaster* of the Indian and Chinese seas. It seems to be even more closely allied to this species than to *P. bismarginatus* and *P. ramphastinus*, Trouess. et Mégn., which are parasites of various South-American birds.

PROCTOPHYLLODES.

Proctophyllodes, Robin et Mégnin, Mém. sur les Sarcoptides plumicoles, p. 539 (1877).

1. Proctophyllodes siallarum, sp. n. (Tab. XXI. figg. 3, *d* ; 4, 4 *a–c*, *♀*.)

d. Corpus minimum, albidum, translucidum, circiter quarta parte minus quam in femina, ad longum, postlateva augrostius quam in fratica, pedes anteriores atque posteriores interstitio longo segregati. Omnes pedes inter se fere aequales, breves. Maxima latitudo corporis ante coxas tertii paris. Ante eas spine recta prope marginem inferna sita est. Pes pedes tertii parte corpus crassitudine diminuit. Appendices anales sat breves, carnosæ, omices, acptratæ lcrators profundæ, in quavis appendice 3 setæ diversæ longitudinis adequi. Disci adhærali in appendicum basi prope marginem interrum laterne siti. Mandibulæ brevis, chelatæ brachia longa, augusta, et ita curva ut chasa apicibus adeos sese attingant.

Long. circiter 0·23–0·8 millim.

♀ adulta. Corpus majus quam in mare. In abdominis spica fures carnosa adest, quæ inferne prope marginem convexum externum in utroque latere spinam lanceolatam sat longam fert. Foree apex rejumpæ lateris in appendicum longum, singatam semilunigam desipit. Epimera primi paris utrimque lateris in media regioni sternali sese conjungunt, secundi paris angulariter retrorsum fracta late disgreta evanuerunt. In corporis latera, paullalam post medium interstitii inter pedes secundii et tertii paris seta est longa inserta est. In vestre media trabecula transversa arcuata seu semicircularis adest, quæ in utroque latere brachio longo angulum seam tertio attingit. Mandibulæ ut in mari.

Long. circiter 0·5 millim.

♀ nympha. Abdomen in appendicum cultrum, carnosam, incisura bipartitam desinit, quæ in superficie externa utrimque lateris spinam lanceolatam sat longam et in spice setam longam gerit.

Larva hexapoda minima, jam rudimenta appendicum analium adultæ præbet.

Ovum longum, fusiforme, translucidum, sat magnum, ad cribri instar perforatum.

Hab. GUATEMALA, Retalhuleu (*Stoll*), in remigibus *Sialia sialis* gregatim.

Male very small, long and narrow, of a whitish, somewhat transparent colour, about one-fourth smaller than the female. The two anterior pairs of legs are separated by a wide interstice from the third pair. The body is broadest at some distance in front of the third pair of coxæ; from the third pair of legs it diminishes in size, and ends in two fleshy appendices, of a short conical shape. The appendices are separated by a deep incision; each of them bears three bristles; the adhesive discs are situated near the inner margin and at the base of the anal lobes. The mandibles are short; the branches of the chelæ are long, slender, and curved, and when closed they touch each other with the points only.

The adult female is a little larger than the male. Its abdomen terminates in a fleshy fork, each point of which bears a lancet-like spine underneath and ends in a narrow, acutely-pointed, somewhat curved blade. The apices of the first pair of epimera of each side touch each other in the sternal region; those of the second pair remain widely separated, and are, at about two-thirds of their length, angularly inflected and directed backward. On the side of the body, somewhat behind the middle of the interstice between the second and third pairs of legs, a long bristle is inserted, and underneath, near the third pair of coxæ, stands, as in the male, a straight, lancet-like spine, which is usual in this genus. A little behind the middle of the ventral surface there is a semicircular transverse trabecula in front of the genital orifice; its branches are continued backward and united with the coxal circles of the fourth pair. The mandibles are as in the male.

In the nymph the abdomen terminates in a conical process, which by a narrow fissure is separated into two lobes, each lobe bearing on its outer edge a lanceolate spine and a long bristle on its top.

The hexapod larva is very small, and shows, though in a rudimentary state, the anal appendices of the adult.

The eggs are long, comparatively large, fusiform, transparent, and cribrated.

NOTE.—Several years have elapsed since the publication of this memoir was commenced in December 1866. In the meanwhile I have become better acquainted with the Acarid-fauna of Europe and some other countries. The clearer insight into the leading features of the geographical distribution in general which I have thus acquired, and of which I have given a résumé in the "Introduction," has made me sceptical with regard to the validity of some of the species described in this work. As an excuse for the synonymical errors into which I may have fallen in some instances, I may be allowed to plead the many difficulties which the study of Acarids offers to the naturalist in a tropical country, in consequence of the extreme delicacy of the soft-bodied species and their great liability to rapid changes of form and colour. Moreover, owing to the unfortunate circumstance that I only brought

G* 2

with me to Europe drawings and more or less fragmentary notes of many of the soft-bodied species, I had no opportunity of comparing the types themselves with European forms and the descriptions of authors.

These remarks will perhaps incite future inquirers to fix their attention more particularly on these fragile forms.

The list which follows contains the names of all the species of Acari hitherto described from Mexico and Central America. In it I have incorporated certain corrections and criticisms, which will be found under the names of the species to which they severally refer.

List of the Species of Acaridea *hitherto described from Mexico and Central America.*

Suborder I. ACARINA-TRACHEATA.

Fam. TROMBIDIDÆ, C. L. Koch.

TROMBIDIUM, Laur.

1. Trombidium mexicanum.

Trombidium mexicanum, Stoll, antea, p. 1, Tab. 1. figg. 1-1 d.

Hab. MEXICO, Presidio.

This undoubtedly bears a remarkable affinity to *T. dubruilli*, A. Dugès; yet it cannot be considered to coincide with that species, as it seems to differ not only in size, but also in some minor details as regards the disposition of the hairs and the form of the palpi and mandibles.

2. Trombidium dubruelli.

Trombidium dubruilli, Alf. Dugès, La Naturaleza, vii. p. 300, L 8. figg. 1-10.

Hab. MEXICO, Guanajuato, Tupataro.

3. Trombidium hispidum.

Trombidium hispidum, Stoll, antea, p. 2, Tab. 11. figg. 1-1 d.

Hab. GUATEMALA, Retalhuleu.

4. Trombidium nasutum.

Trombidium nasutum, Stoll, antea, p. 2, Tab. III. figg. 1-1 g.

Hab. GUATEMALA, Retalhuleu.

5. Trombidium quinque-maculatum.

Trombidium quinque-maculatum, Stoll, antch, p. 3, Tab. IV. figg. 1–1 c.

Hab. GUATEMALA, near the city.

6. Trombidium muricola.

Trombidium muricola, Stoll, antch, p. 5, Tab. II. figg. 3–3 d.
Trombidium guayaricola, Stoll, antch, p. 4, Tab. II. figg. 3–3 c.

Hab. GUATEMALA, Retalhuleu, Antigua.

T. guayaricola and *T. muricola* are, I believe, only colour-varieties of one and the same species. I was led into error by the different habitats, and by the discrepancies between my original drawings, which were made, at a year's interval, at different places, Retalhuleu and Antigua. I propose to drop the name of *T. guayaricola* and to keep only that of *T. muricola*.

7. Trombidium trilineatum.

Trombidium trilineatum, Stoll, antch, p. 4, Tab. I. figg. 3–3 c.

Hab. GUATEMALA, Antigua.

8. Trombidium albicolle.

Trombidium albicolle, Stoll, antch, p. 6, Tab. I. figg. 6, 3 d.

Hab. GUATEMALA, Antigua.

RHYNCHOLOPHUS, Dugès.

1. Rhyncholophus erinaceus.

Rhyncholophus erinaceus, Stoll, antch, p. 6, Tab. IV. figg. 2–2 d.

Hab. GUATEMALA, Antigua.

Fam. ACTINEDIDÆ.

ACTINEDA, C. L. Koch.

1. Actineda baocarum.

Acarus baccarum, Linn. Syst. Nat. 10th edit. p. 617.
Acarus vitis, Schrank, Enum. Ins. Austr. indig. p. 519.
Trombidium corrigerum, Herm. Mém. Aptérolog. p. 38, t. 2. fig. 9.
Actineda furvata, Stoll, antch, p. 7, Tab. V. figg. 1, 1 c.
Actineda antiguensis, Stoll, antch, p. 7, Tab. V. figg. 3–3 d.
Actineda rételfera, Stoll, antch, p. 7, Tab. V. figg. 3–3 c.

Hab. EUROPE.—GUATEMALA, Retalhuleu, Antigua.—SOUTH AMERICA.

After having studied more attentively the European *A. baccarum*, L. (=*cornigerum*, Herm.), I believe that the above-named Guatemalan forms are but varieties of colour and age of this species, as I cannot find any tangible differences between my original drawings and European specimens, except those of size and colour, which in *Actineda* are of no value. This view is confirmed by the circumstance that Berlese has identified preserved specimens of *Actineda* from Rio Apa (Paraguay) and from Buenos Ayres with the European species, with the remark "parum ab Europæa diversa"*. It therefore seems that *A. baccarum*, L. (=*A. vitis*, Schrank, = *Tromb. cornigerum*, Hermann), is one of those fundamental and characteristic types which occupy an extremely extensive geographical area, a fact I have discussed at length in the "Introduction" to the present memoir.

Fam. TETRANYCHIDÆ, Kramer.

TETRANYCHUS, Dufour.

1. Tetranychus guatemalæ-novæ.

Tetranychus guatemalæ-novæ, Stoll, antea, p. 8, Tab. VI. figg. 1-1 c.

Hab. GUATEMALA, near the city.

2. Tetranychus dugesi.

Tetranychus dugesi, D. Cano y Alcacio, La Naturaleza, vii. p. 197, t. 2. figg. 1-3.

Hab. MEXICO. On *Medicago denticulata*, Willd.

Fam. HYDRACHNIDÆ, Neuman.

The corrections in the synonymy of the species of this family described by me are made on the competent authority of Herr Koenike of Bremen, who, at my request, has favoured me with his opinion on them.

ATAX, Fabr.

1. Atax alticola.

Atax alticola, Stoll, antea, p. 9, Tab. VII. figg. 1-1 g.
Atax septem-maculatus, Stoll, antea, p. 9, Tab. VIII. figg. 1-1 c.

* A. Berlese (Acari Austro-Americani, p. 12) identifies his American specimens with *Acarus vitis*, Schrank, which he considers to be synonymous with *Trombidium cornigerum*, Hermann. But as Schrank himself, in his description of *A. vitis* (Enum. Ins. Austr. indig. p. 519. no. 1067), says that it differs from *A. baccarum*, L., only by the disposition of the hairs on the legs, I think that we must regard *A. vitis*, Schrank, and *Acarus baccarum*, L., as belonging to one and the same species.

Atax septem-maculatus, var. *ypsilon*, Stoll, anteà. p. 10, Tab. IX. figg. 1-1 *e*.

Hab. GUATEMALA, near the city.

According to Herr Koenike, *A. septem-maculatus* and its var. *ypsilon* are probably nymphal stages of *A. alticola*.

2. Atax dentipalpis.

Atax dentipalpis, Stoll, antcà, p. 10, Tab. X. figg. 1-1 *d*.

Hab. GUATEMALA, near the city.

This, according to the same authority, should be referred to the widely distributed *A. crassipes*, O. F. Müll.

NESÆA, Neuman.

1. Nesæa alzatei.

Atax alzatei, A. Dugès, La Naturaleza, vi. p. 344, t. 8. figg. 1-19.

Hab. MEXICO, Guanajuato.

This species has been ascribed by its author to the genus *Atax*.

2. Nesæa guatemalensis.

Nesæa guatemalensis, Stoll, antcà, p. 11, Tab. X. figg. 2-3 *b* (♀), and Tab. XI. figg. 1-1 *f* (♂).

Hab. GUATEMALA, near the city.

3. Nesæa numulus.

Nesæa numulus, Stoll, antcà, p. 13, Tab. XI. figg. 2-2 *e*.

Hab. GUATEMALA, near the city.

LIMNESIA, Neuman.

1. Limnesia guatemalteca.

Limnesia guatemalteca, Stoll, antcà, p. 13, Tab. VII. figg. 2-2 *e*.

Hab. GUATEMALA, near the city.

From the form of the palpi, Herr Koenike considers this to be a nymph, probably of the following species.

2. Limnesia longipalpis.

Limnesia longipalpis, Stoll, antcà, p. 13, Tab. IX. figg. 2-2 *e*.

Hab. GUATEMALA, near the city.

3. Limnesia putearum.

Limnesia putearum, Stoll, antch, p. 14, Tab. VII. figg. 3-3 c.

Hab. GUATEMALA, Antigua.

4. Limnesia læta.

Limnesia læta, Stoll, antch, p. 14, Tab. VIII. figg. 2-2 d.

Hab. GUATEMALA, near the city.

Fam. BDELLIDÆ.

BDELLA, Latr.

1. Bdella splendida.

Bdella splendida, Stoll, antch, p. 15, Tab. III. figg. 2-2 c.

Hab. GUATEMALA, near the city.

Fam. EUPODIDÆ.

SCYPHIUS, Koch.

1. Scyphius maniacus.

Scyphius maniacus, Stoll, antch, p. 17, Tab. VI. figg. 2-3 d.

Hab. GUATEMALA, Retalhuleu.

Fam. IXODIDÆ.

ARGAS, Latr.

1. Argas talaje.

Argas talaje, Guérin-Méneville, Revue et Mag. Zool. 1849, p. 342, t. 9.

Hab. GUATEMALA, Casas Viejas de Guastatoya.

2. Argas turicata.

Argas turicata, A. Dugès, La Naturaleza, vi. p. 195, t. 4 c. figg. a 1-8.

Hab. MEXICO, Guanajuato.

A parasite on hogs.

3. Argas megnini.

Argas megnini, A. Dugès, La Naturaleza, vi. p. 197, t. 4 c. figg. a 1-8.

Hab. MEXICO, Guanajuato.

This species is stated to attach itself to the skin in the ears of horses, cattle, and even man.

ORNITHODOROS, Koch.

1. Ornithodorus coriaceus.

Ornithodorus coriaceus, C. L. Koch, Arch. f. Naturg. x. Bd. 1, p. 219, no. 1 ; Übers. d. Arachniden-syst. Heft iv. pp. 12, 31, t. 1. figg. 2, 3.

Hab. MEXICO (*teste Koch*).

Berlese, in his 'Acari Austro-Americani,' p. 25, mentions *Ornithodoros coriaceus*, Koch, from Rio Apa (Paraguay), but without entering into details.

IXODES, C. L. Koch.

1. Ixodes boarum.

Ixodes boarum, Stoll, antea, p. 18, Tab. XIII. figg. 1-1 e, and Tab. XIV. fig. 4.

Hab. GUATEMALA, Retalhuleu.

2. Ixodes pygmæus.

Ixodes pygmæus, C. L. Koch, Arch. f. Naturg. x. Bd. 1, p. 233, no. 18 ; Übers. d. Arachnidensyst. Heft iv. pp. 29, 107, and t. 27. figg. 81 a, b (♀).

Hab. MEXICO.—BRAZIL (*teste C. L. Koch*).

AMBLYOMMA, C. L. Koch.

1. Amblyomma mixtum.

Amblyomma mixtum, C. L. Koch, Arch. f. Naturg. x. Bd. 1, p. 227, no. 17 ; Übers. d. Arachniden-syst. Heft iv. pp. 74, 75, t. 13. figg. 47 (♂), 48 (♀); Stoll, antea, p. 19, Tab. XII. figg. 1-1 i (♀), 2-2 b (♂).

Hab. MEXICO (*teste C. L. Koch*); GUATEMALA, Retalhuleu, Antigua ; NICARAGUA, Chontales ; COSTA RICA, Caché.

2. Amblyomma dissimile.

Amblyomma dissimile, C. L. Koch, Arch. f. Naturg. x. Bd. 1, p. 225, nn. 10 (♂ ♀); Übers. d. Arachnidensyst. Heft iv. pp. 17, 64, 66, t. 12. figg. 37 (♂), 38 (♀).

Hab. MEXICO (*teste Koch*).

3. Amblyomma tenellum.

Amblyomma tenellum, C. L. Koch, Arch. f. Naturg. x. Bd. 1, p. 227, no. 16 (♂); Übers. d. Arachnidensyst. Heft iv. pp. 17, 76, 79, t. 14. fig. 51 (♂).

Hab. MEXICO (*teste Koch*).

4. Amblyomma ovale.

Amblyomma ovale, C. L. Koch, Arch. f. Naturg. x. Bd. 1, p. 227, no. 20 ; Übers. d. Arachniden-syst. Heft iv. pp. 18, 79, 80, t. 14. fig. 52 (♂).

Hab. MEXICO (*teste Koch*).

5. **Amblyomma forelI.**

Amblyomma foreli, Stoll, antrb, p. 31, Tab. XII. figg. 3-3 *b*, and Tab. XIV. figg. 3-3 *d* (♀).

Hab. GUATEMALA, Retalhulen.

6. **Amblyomma crassipunctatum.**

Amblyomma crassipunctatum, Stoll, antch, p. 22, Tab. XIV. figg. 1-1 *b* (*d*).

Hab. NICARAGUA, Chontales.

7. **Amblyomma sabanerœ.**

Amblyomma sabanero, Stoll, antch, p. 22, Tab. XIV. figg. 2-2 *i* (♀).

Hab. GUATEMALA, near Retalhulen.

Fam. ORIBATIDÆ.

Subfam. *PTEROGASTERINÆ.*

ORIBATA, Latr.

1. **Oribata centro-americana.**

Oribata centro-americana, Stoll, antch, p. 21, Tab. XV. figg. 1-1 *f*.

Hab. BRITISH HONDURAS, R. Llondo, R. Sarstoon, Belize; GUATEMALA, Antigua, Guatemala city; PANAMA, Volcan de Chiriqui.

2. **Oribata rugifrons.**

Oribata rugifrons, Stoll, antch, p. 23, Tab. XV. figg. 2-3 *d* (? nymph, Tab. XV. figg. 3-3 *d*).

Hab. BRITISH HONDURAS, Belize; GUATEMALA, Retalhulen.

Subfam. *APTEROGASTERINÆ.*

HOPLOPHORA, C. L. Koch.

1. **Hoplophora retaltecn.**

Hoplophora retaltecn, Stoll, antch, p. 27, Tab. XV. figg. 4-1 *f*.

Hab. GUATEMALA, Retalhulen.

Fam. NICOLETIELLIDÆ.

NICOLETIELLA, R. Canestrini.

1. **Nicoletialla neotropica.**

Nicoletiella neotropica, Stoll, antch, p. 27, Tab. XVI. figg. 1-1 *c*.

Hab. GUATEMALA, Retalhulen.

Fam. GAMASIDÆ.

Subfam. UROPODINÆ.

UROPODA, Latr.

1. Uropoda mœsta.

Uropoda mœsta, Walck. & Gervais, Hist. nat. des Ins. Aptères, iii. p. 221, t. 34. fig. 5.

Hab. MEXICO. Found as a parasite on *Polydesmus mexicanus.*

2. Uropoda echinata.

Uropoda echinata, Stoll, antek, p. 28, Tab. XVI. figg. 2–2 c.

Hab. GUATEMALA, Antigua.

3. Uropoda inæquipunctata.

Uropoda inæquipunctata, Stoll, antek, p. 29, Tab. XVI. figg. 3–3 d.

Hab. GUATEMALA, Retalhuleu.

4. Uropoda discus.

Uropoda discus, Stoll, antek, p. 29, Tab. XVII. figg. 4–4 c.

Hab. GUATEMALA, Retalhuleu.

5. Uropoda centro-americana.

Uropoda centro-americana, Stoll, antek, p. 30, Tab. XVII. figg. 1–1 f, and 2–2 b (nymph).

Hab. NICARAGUA, Chontales.

6. Uropoda piriformis.

Uropoda piriformis, Stoll, antek, p. 31, Tab. XVII. figg. 5–3 d.

Hab. MEXICO, Jalapa.

Subfam. GAMASINÆ.

MEGISTHANUS, T. Thorell.

1. Megisthanus gigantodes.

Megisthanus gigantodes, Stoll, antek, p. 31, Tab. XVIII. figg. 1–1 d (♀), 2–2 g (♂).

Hab. GUATEMALA, Accitono, Guatemala city, Zapote, Retalhuleu, Cholbuitz.

2. Megisthanus armiger.

Megisthanus armiger, A. Berlese, Bull. Soc. Ent. Ital. xx. p. 204, t. 9, iv. fig. 1; Stoll, antek, p. 34, Tab. XIX. figg. 1–1 e (♂).

Hab. MEXICO, Jalapa.—PARAGUAY, Rio Apa.

M* 2

CELÆNOPSIS, Berlese.

1. Celænopsis uropodoides.

Celænopsis uropodoides, Stoll, antea, p. 35, Tab. XVI. figg. 4–4 d, Tab. XIX. figg. 3–3 b (♀).

Hab. BRITISH HONDURAS, R. Hondo.

2. Celænopsis magisthanoides.

Celænopsis magisthanoides, Stoll, antea, p. 36, Tab. XIX. fig. 4, Tab. XX. figg. 1–1 b (♀)

Hab. PANAMA, Bugaba.

PACHYLÆLAPS, Berlese.

1. Pachylælaps hærus.

Pachylælaps hærus, var. *mexicanus*, Stoll, antea, pp. 37, 38, Tab. XIX. figg. 2–2 e,
Pachylælaps hærus, Berlese, Bull. Soc. Ent. Ital. xx. p. 196, t. 6. fig. 3.

Hab. MEXICO, San Andres Tuxtla. [Typus: Matto Grosso, Brazil.]

HOLOSTASPIS, Kolenati.

1. Holostaspis marginatus.

Holostaspis marginatus (Hermann), Stoll, antea, p. 39, Tab. XX. figg. 2–2 d.

Hab. GUATEMALA, Retalhuleu; NICARAGUA, Chontales.—SOUTH AMERICA, Brazil, La Plata, Paraguay.

Suborder II. ACARINA-ATRACHEATA.

Fam. SARCOPTIDÆ.

Subfam. *ANALGESINÆ*.

MEGNINIA, Berlese.

1. Megninia pteroglossorum.

Megninia pteroglossorum, Stoll, antea, p. 40, Tab. XXI. figg. 5–5 b.

Hab. GUATEMALA, Retalhuleu.

PTEROLICHUS, Robin et Mégnin.

1. Pterolichus momotorum.

Pterolichus momotorum, Stoll, antea, p. 41, Tab. XXI. figg. 1–1 b (♂), 2, 3 e (♀).

Hab. GUATEMALA, Retalhuleu.

PROCTOPHYLLODES, Robin et Mégnin.

1. Proctophyllodes glabarum.

Proctophyllodes glabarum, Stoll, antea, p. 42, Tab. XXI. fgg. 5 (♂), 6–6 c (♀).

Hab. GUATEMALA, Retalhuleo.

N.B.—I have noticed the presence of a few other genera in Guatemala, but my notes and drawings respecting them are too fragmentary to allow specific determination. These genera are: *Linopodes*, C. L. Koch, *Gamasus*, Latr. (sensu stricto), *Pteroptus*, Duf., *Tyroglyphus*, Latr.

INDEX.

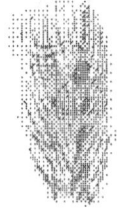

1a

1

1d

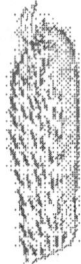

2a

1b

1c

3a

2b

2c

3.

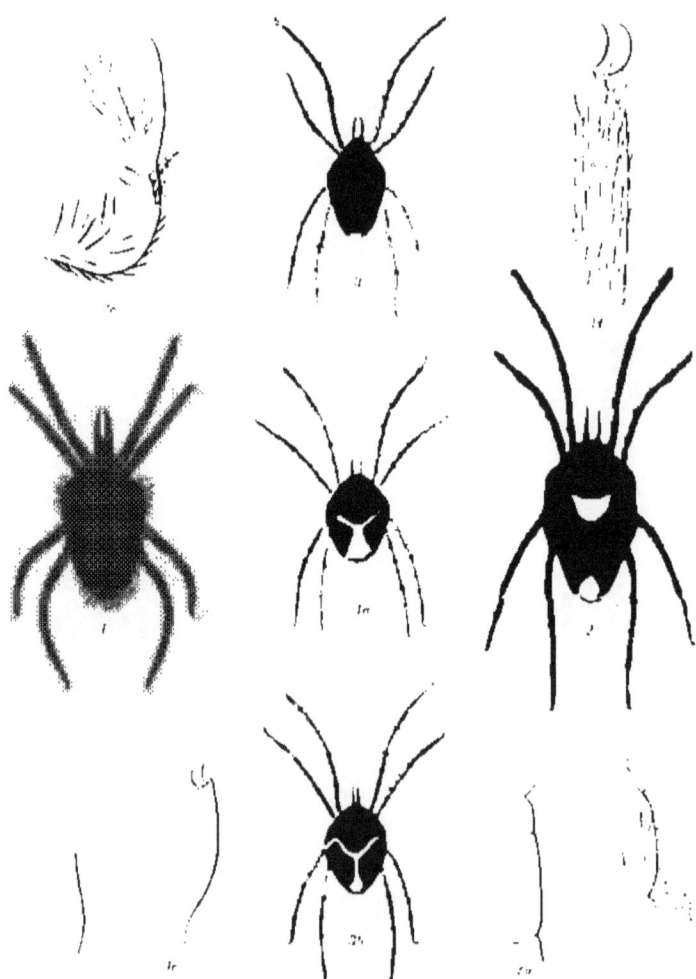

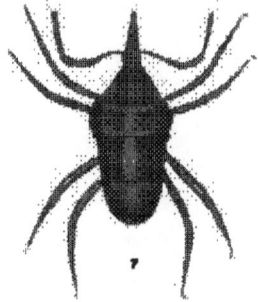

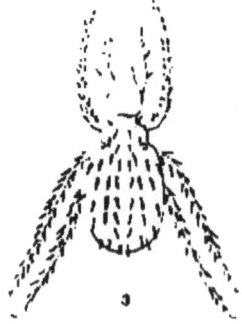

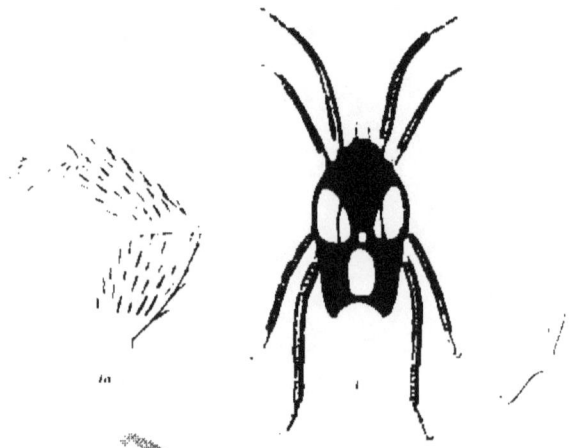

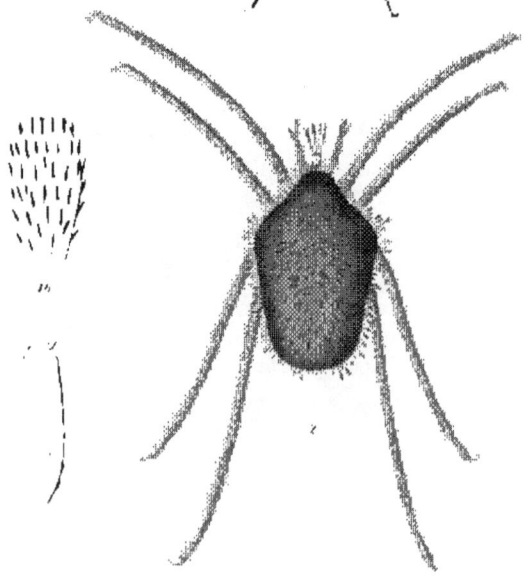

2a *1a* *2b*

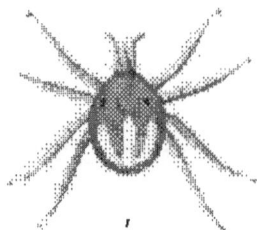

1 *1a* *c*

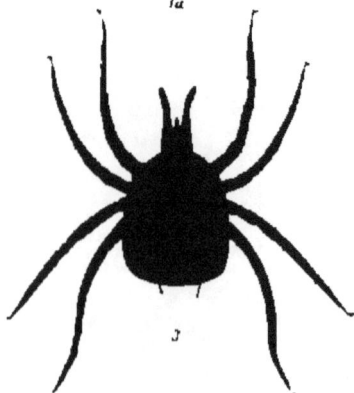

2c *3b*

2d *a* *3* *3c*

1a.

1.

1c.

1b.

2a.

2.

2b.

2c.

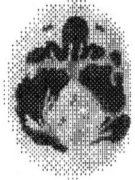

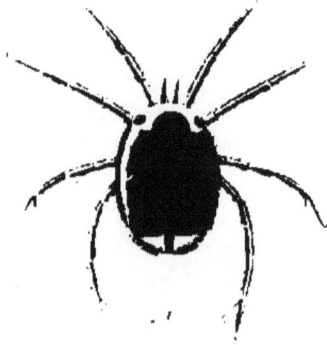

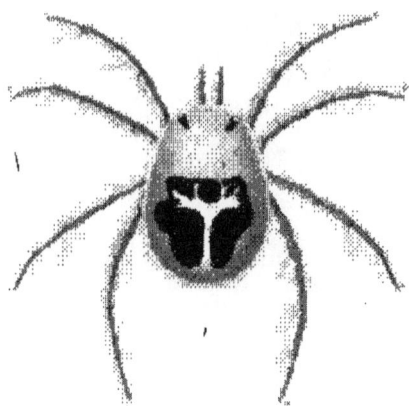

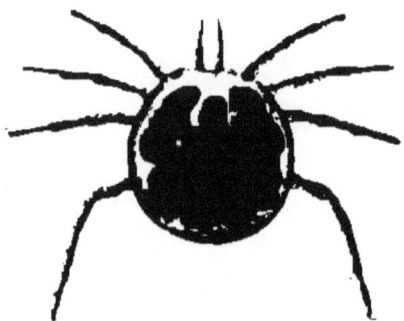

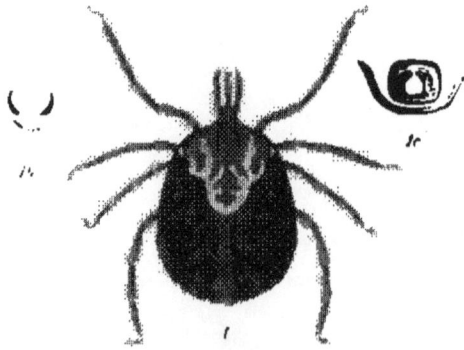

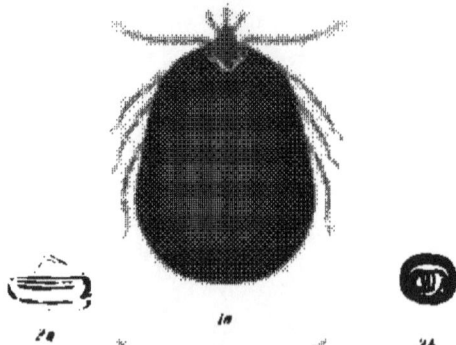

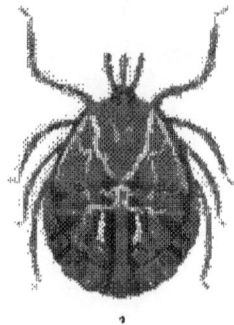

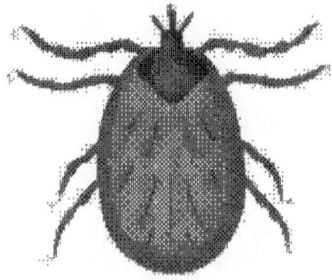

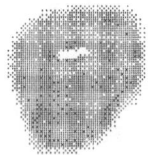

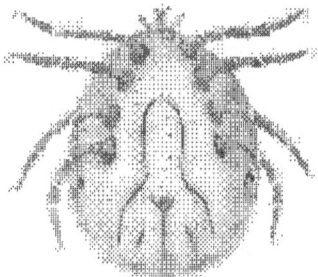

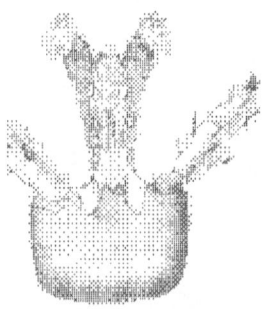

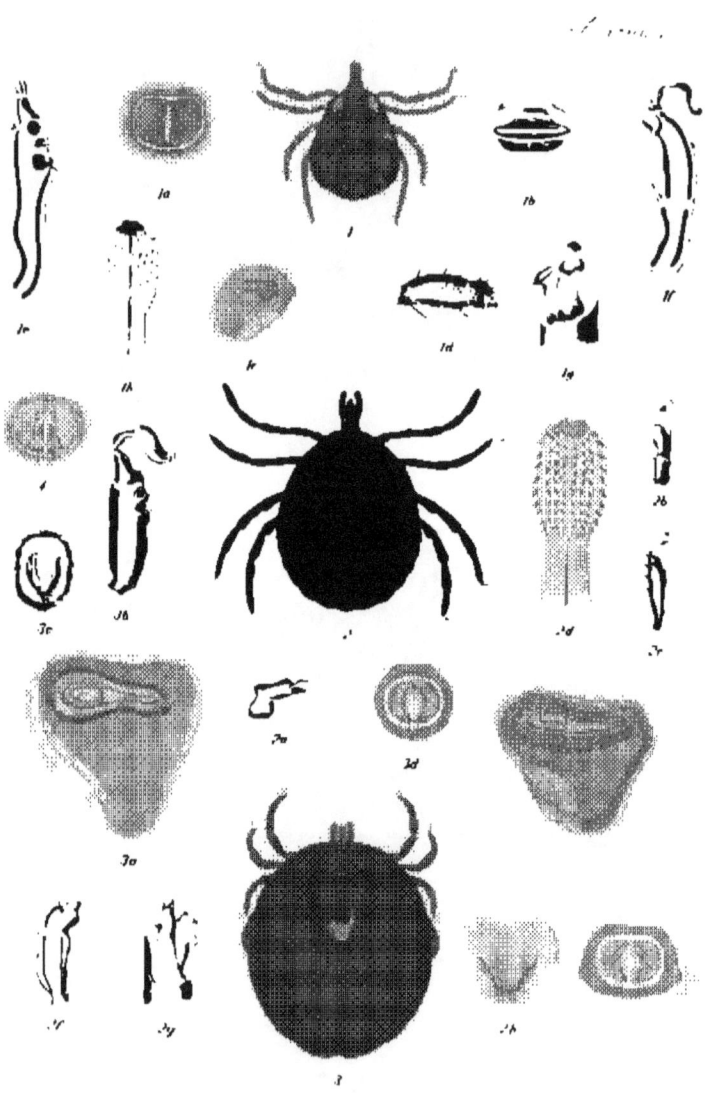

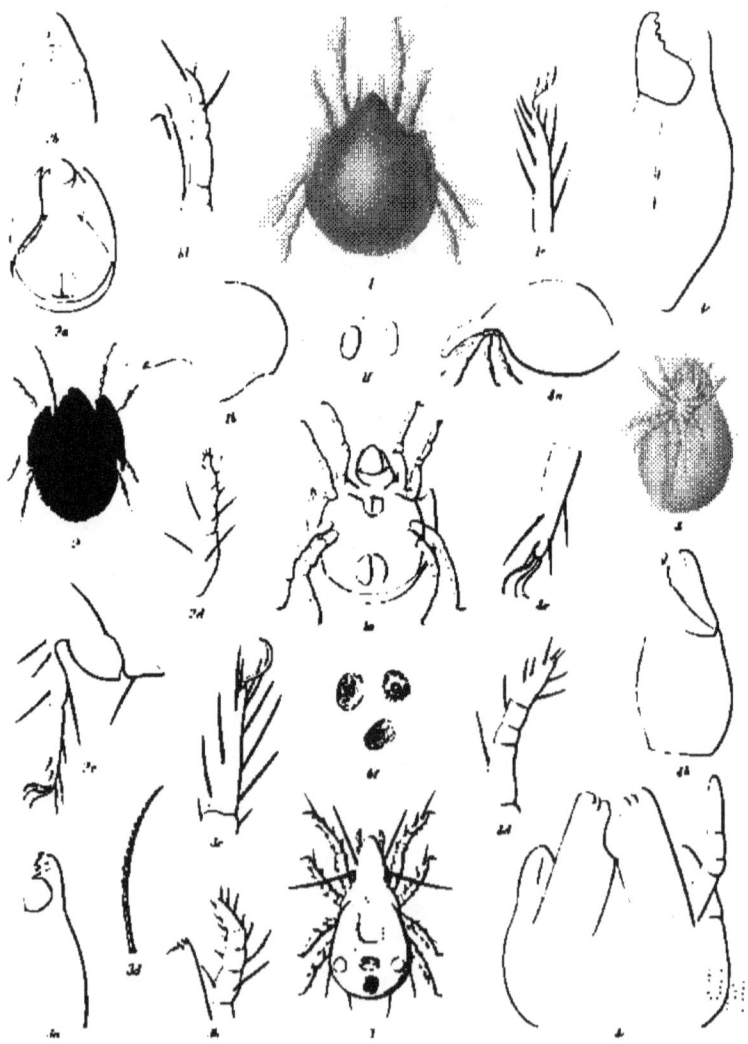

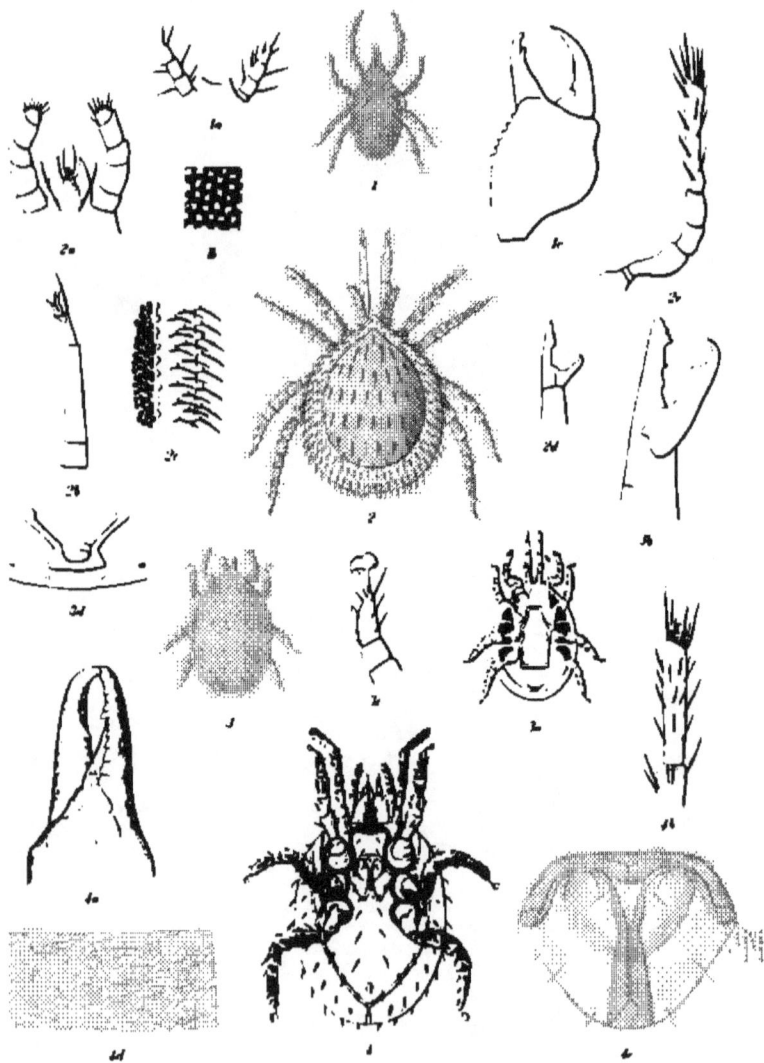

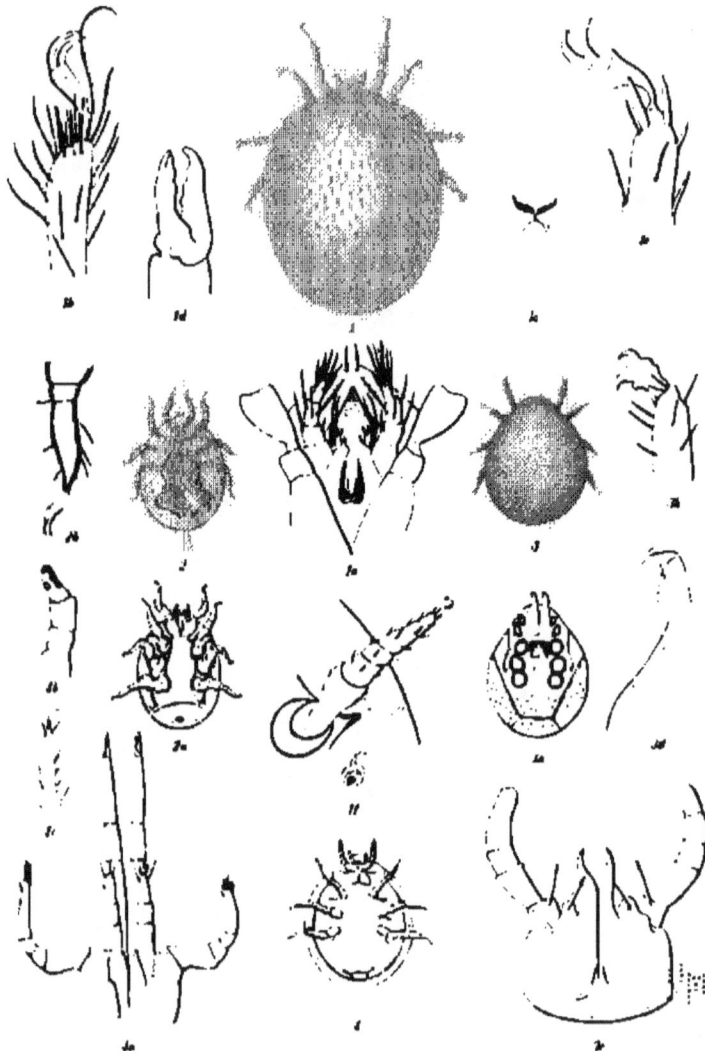

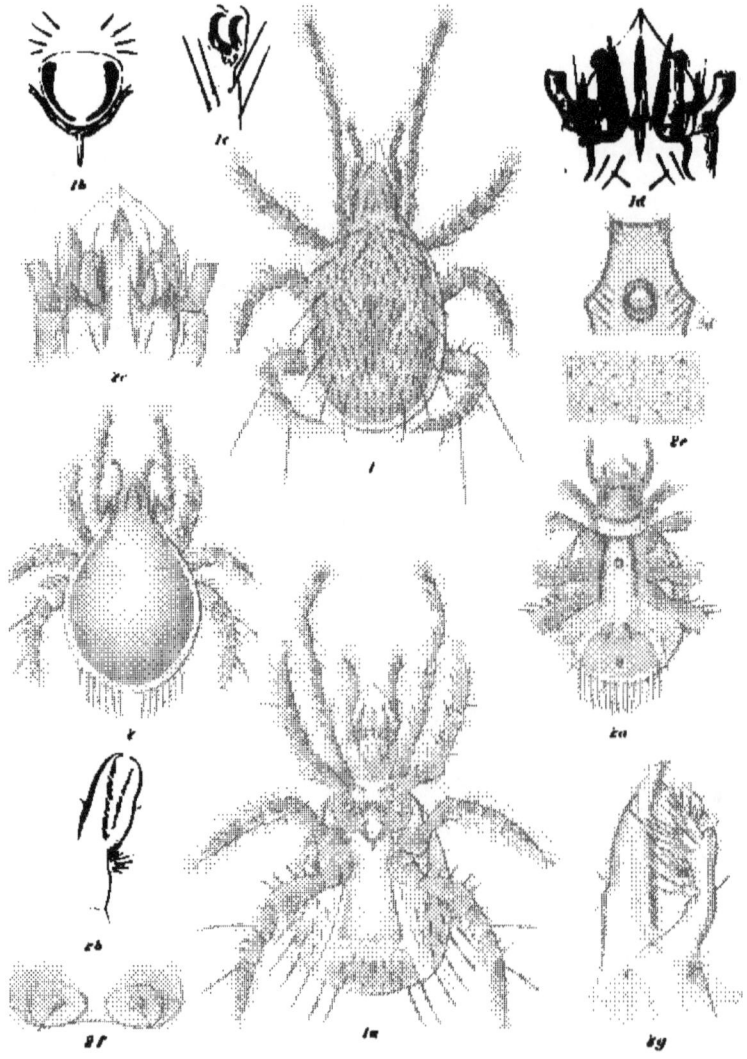

NEGRITHANIS GIGANTODES.

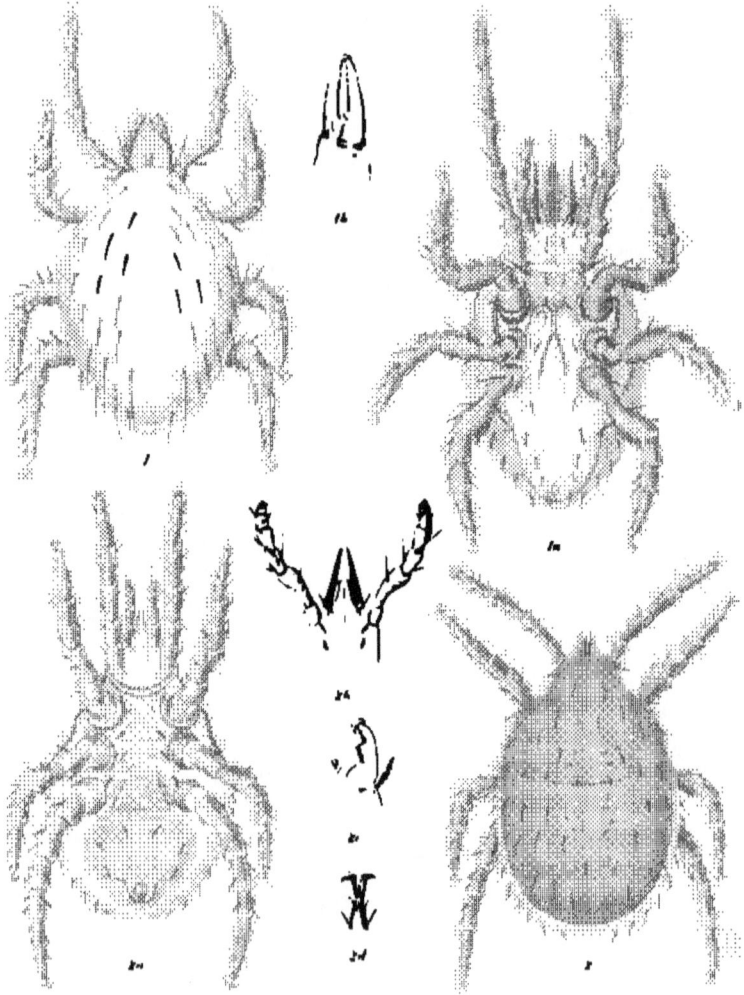

1, 1a, b, c. OPLONOPSIS MEGISTHENODES
2, 2a, b, c, d. HOLOSTASPIS MARGINATUS.

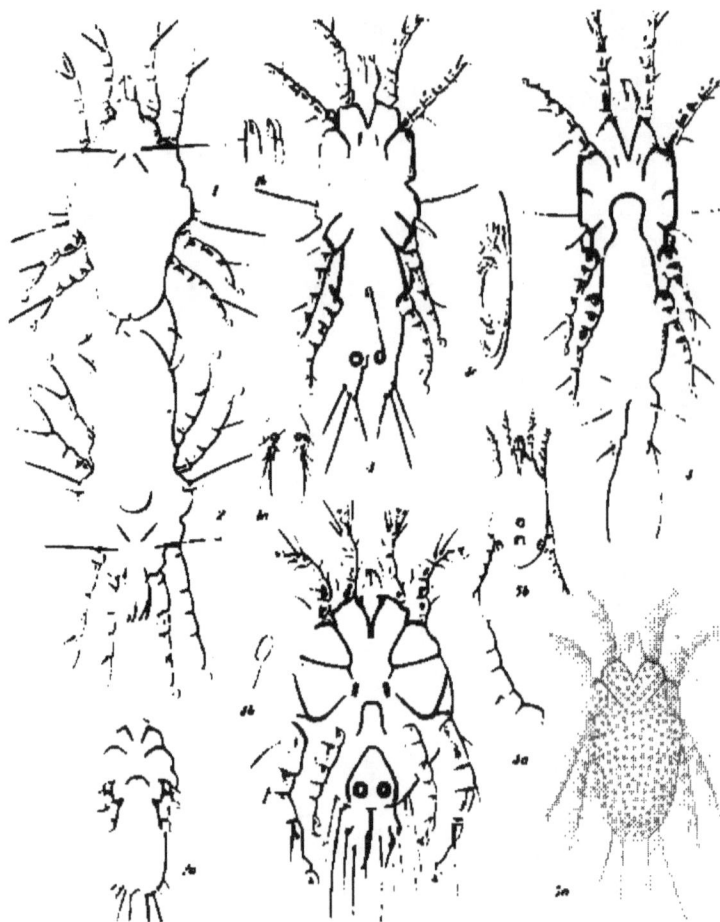

www.ingramcontent.com/pod-product-compliance
Lightning Source LLC
Chambersburg PA
CBHW021823190326
41518CB00007B/721